Analog Circuits and Signal Processing

More information about this series at http://www.springer.com/series/7381

Nele Reynders • Wim Dehaene

Ultra-Low-Voltage Design of Energy-Efficient Digital Circuits

Springer

Nele Reynders
ESAT-MICAS, KU Leuven
Heverlee, Belgium

Wim Dehaene
ESAT-MICAS, KU Leuven
Heverlee, Belgium

ISSN 1872-082X ISSN 2197-1854 (electronic)
Analog Circuits and Signal Processing
ISBN 978-3-319-16135-8 ISBN 978-3-319-16136-5 (eBook)
DOI 10.1007/978-3-319-16136-5

Library of Congress Control Number: 2015935431

Springer Cham Heidelberg New York Dordrecht London

Printed on acid-free paper

Springer International Publishing AG Switzerland is part of Springer Science+Business Media (www.springer.com)

Preface

These days the Internet of Things (IoT) is the big focus when conceiving digital systems. Given the billions of nodes that are ultimately projected for IoT, low energy signal processing is the holy grail. Ignoring the question whether or not the grail can actually be found, we thus embarked on a quest for a significant reduction of the energy consumption per digital operation. The first parameter that catches the eye when wishing to reduce energy is the power supply voltage. We all know that dynamic energy per operation is "cap-times-V_{dd}-square". So the obvious conclusion is: let us reduce the power supply voltage as much as possible. That was the starting point of the design.

As always in research, it is not that simple: the square is only dynamic energy, at low voltages the robustness reduces considerably and so on. Where this research journey will end, is the subject of the book you are holding. However, we can already tell you that it was an exciting journey. Among others, transmission gate logic was reborn and general purpose technologies proved to be more low power than actual low power technologies. The temporary end of the quest, the result of one Ph.D., is described in this book. It consists of novel circuits with lower than state-of-the-art energy consumption, at least at the time of this writing. These circuits, ranging from a small adder to a complete JPEG encoder, are proven on silicon in different technologies. Our research results will give you the practical inspiration to design ultra-low-energy circuits based on the novel principles we describe. There is no fully automated design flow yet. That is for a next book.

We hope, dear reader, that after reading this book you will share our enthusiasm. We are convinced that we have set important steps toward true low energy design. We actually show that mismatch can be dealt with despite the sub-threshold exponential current regime. As such we can guarantee sub-threshold robustness at ultra-low supply voltage. Enjoy!

Heverlee, Belgium
January 2015

Nele Reynders
Wim Dehaene

Contents

1 Introduction 1
1.1 Supply Voltage Reduction: A Brief History 1
1.2 Reducing the Energy Consumption 3
1.2.1 Definitions 4
1.2.2 Minimizing the Energy Consumption 5
1.2.3 Energy-Delay Product 7
1.3 Ultra-Low-Voltage Digital Design 8
1.4 Applications 10
1.5 Current State-of-the-Art in Literature 10
1.6 Outline of This Work 15
References 17

2 Sub-Threshold Operation: Theory and Challenges 19
2.1 Transistor Operation 20
2.1.1 Different Operating Regions 21
2.1.2 Threshold Voltage 24
2.1.3 Region of Interest 28
2.2 Challenges of Sub-Threshold Operation 29
2.2.1 Performance 29
2.2.2 Leakage 30
2.2.3 Variability 31
2.2.4 Temperature 34
2.3 Technology Scaling 35
2.3.1 Fundamental Limits 36
2.3.2 Impact of Scaling 40
2.3.3 Model Accuracy 41
2.4 Transistor Type 42
2.5 Conclusion 43
References 43

3 Gate-Level Building Blocks 47
3.1 Circuit Topology Comparison 48
3.1.1 Standard CMOS Logic 48
3.1.2 Pseudo-nMOS Logic 60
3.1.3 Pass Transistor Logic 62
3.1.4 Transmission Gate Logic 64
3.1.5 Other Topologies 71
3.2 Chosen Circuit Topologies 75
3.2.1 Logic Gates 75
3.2.2 Inverter 76
3.3 Memory Elements 77
3.3.1 Latch 78
3.3.2 Flip-Flop 80
3.4 Sizing in Different Prototypes 80
3.5 Conclusion 80
References 81

4 Architectural Design 85
4.1 Theoretical Considerations 86
4.1.1 Energy Ratio 87
4.1.2 Total Energy Consumption 88
4.2 Cascading Logic Gates 92
4.2.1 Concept 92
4.2.2 Trade-Off 94
4.2.3 Differential TG Logic 96
4.2.4 Realization 98
4.3 Pipelining 98
4.3.1 Concept 98
4.3.2 Benefits and Drawbacks 99
4.3.3 Pipelining Schemes 100
4.3.4 Design Considerations 105
4.4 Design Methodology 107
4.4.1 Design 107
4.4.2 Layout 108
4.5 I/O Circuits 109
4.6 Conclusion 111
References 112

5 Datapath Blocks 113
5.1 Adder 114
5.1.1 Proof of Concept 114
5.1.2 Architecture 114
5.1.3 Ultra-Low-Voltage Design 114
5.1.4 Measurement Results 117
5.1.5 State-of-the-Art Comparison 119
5.1.6 Conclusion 120

5.2 Multiply-Accumulate Unit 120
5.2.1 Proof of Concept 120
5.2.2 Architecture 121
5.2.3 Ultra-Low-Voltage Design 123
5.2.4 Measurement Results 130
5.2.5 State-of-the-Art Comparison 136
5.2.6 Conclusion 137
5.3 Conclusion 138
References 138

6 JPEG Encoder 141
6.1 Proof of Concept 142
6.2 JPEG Encoding Algorithm 142
6.3 Ultra-Low-Voltage Design 144
6.4 Implementation 144
6.4.1 Timing 144
6.4.2 2D-DCT 146
6.4.3 Quantization 148
6.4.4 Zigzag Matrix and Huffman Encoder 149
6.4.5 Lookup Tables 156
6.5 Measurement Results 159
6.6 State-of-the-Art Comparison 161
6.7 Lookup Table Improvements 165
6.8 Conclusion 169
References 169

7 Conclusion 171
7.1 General Conclusions 171
7.2 State-of-the-Art Comparison 174
7.3 Main Contributions 177
7.4 Suggestions for Future Work 178
7.4.1 Energy-Efficient SRAM 179
7.4.2 Other Technologies 179
7.4.3 Standard Digital Design Flow 179
7.4.4 Inter-Die Variations 180
7.4.5 Temperature-Dependence 180
7.4.6 Efficient DC-DC Converter 180
Reference 180

A Current State-of-the-Art in Literature 181
References 184

Index 189

Abstract

Nowadays, energy-efficiency is becoming more and more a decisive parameter for digital systems, driven by the ever increasing number of portable applications. Mobile phones are an obvious example, but many other portable electronic devices are emerging which have less stringent speed requirements but even more critical energy requirements. Since their stand-alone time is dependent on the fixed available energy budget, research toward significant improvements in energy consumption per operation is paramount. Especially medical applications such as biomedical sensor nodes can benefit greatly from a drastically increased energy-efficiency.

By extremely reducing the supply voltage of digital CMOS circuits, their dynamic energy consumption decreases quadratically. Therefore, operating digital systems at ultra-low supply voltages can result in significant energy savings. However, ultra-low-voltage circuits pose many challenges as well. The current decreases exponentially, causing the delay to increase considerably. Hence, inherently, it is only possible to achieve low to moderate circuit performance. Circuits operating at such low supply voltages are much more sensitive to variations, which can severely compromise the yield. Moreover, the decreased current ratios pose a threat to reliable functionality of the circuits.

This book aims to design ultra-low-voltage digital circuits which are not only energy-efficient, but also provide answers to these various challenges. A focus is given toward designing variation-resilient circuits, as this is key to guarantee high yield. Additionally, operating frequencies of $n \times 10\,\text{MHz}$ are targeted to establish ultra-low-voltage systems as an attractive option for industrial applications.

To accomplish these various research aims, careful attention must be paid to all abstraction levels of digital design. Therefore, a complete design methodology is presented, which follows a bottom-up approach from transistor-level circuit design up to architecture-level recommendations. This ultra-low-voltage design strategy is generally applicable for all types of signal processing applications. This is demonstrated by four implemented prototypes: three datapath elements, i.e. a logarithmic adder and two multiply accumulate units, and a full JPEG encoder in 90 nm and 40 nm CMOS technologies. These prototypes have successfully obtained the predefined research goals and have thereby succeeded to effectively validate the proposed design methodology.

Acronyms

BAN	Body area network
CDF	Cumulative distribution function
CEF	Constant electric field
CMOS	Complementary metal-oxide-semiconductor
CPL	Complementary pass transistor logic
CV	Constant voltage
DCT	Discrete cosine transform
DIBL	Drain-induced barrier lowering
DPG	Datapath generator
DSP	Digital signal processor
ECG	Electrocardiogram
ECRL	Efficient charge recovery logic
EDP	Energy-delay product
EOB	End-of-block
FBB	Forward body biasing
FFT	Fast fourier transform
FIR	Finite impulse response
FOM	Figure of merit
FSM	Finite state machine
HVT	High threshold voltage
INWE	Inverse narrow width effect
I/O	Input/output
JPEG	Joint photographic experts group
LVT	Low threshold voltage
MAC	Multiply-accumulate unit
MC	Monte carlo
MEP	Minimum-energy point

MSB	Most significant bit
NM	Noise margin
nMOS	n-channel MOS transistor
NOCG	Non-overlapping clock generator
PDF	Probability density function
PDN	Pull-down network
PDP	Power-delay product
PFAL	Positive feedback adiabatic logic
pMOS	p-channel MOS transistor
PTM	Predictive technology model
PUN	Pull-up network
RBB	Reverse body biasing
RFID	Radio-frequency identification
RSCE	Reverse short-channel effect
SAPTL	Sense amplifier-based pass transistor logic
SCE	Short-channel effect
SOI	Silicon-on-insulator
SRAM	Static random access memory
STSCL	Sub-threshold source-coupled logic
SVT	Standard threshold voltage
TG	Transmission gate
VTC	Voltage transfer characteristic
ZRL	Zero runlength

Symbols

A_{V_T}	Pelgrom coefficient of the threshold voltage
C	Capacitance
C_D	Depletion layer capacitance
C_{ox}	Gate oxide capacitance per unit area
E_{dyn}	Dynamic energy
E_{stat}	Static energy
E_{tot}	Total energy
f_{clk}	Clock frequency
F_v	Variation factor
I_0	Transistor current when $V_{gs} = V_T$
I_{ds}	Drain-source current of a transistor
I_{leak}	Leakage current
I_{off}	Off-current of a transistor
I_{on}	On-current of a transistor
k	Boltzmann constant ($= 1.380650524 \cdot 10^{-23}$J/K)
K_i	Relative dielectric constant of a material i
L	Length of a transistor
n	Technology-dependent parameter
N_A	Doping concentration of the substrate
n_i	Intrinsic carrier concentration
$NM_{L/H}$	Low / high noise margin
P_{dyn}	Dynamic power
P_{stat}	Static power
P_{tot}	Total power
q	Electric charge of an electron ($= 1.602176565 \cdot 10^{-19}$C)

S_S	Sub-threshold slope
T	Absolute temperature
t_{clk}	Clock cycle
t_d	Gate delay
t_f	Fall time
t_{ox}	Oxide thickness
t_p	Propagation delay
t_{pHL}	High-to-low propagation delay
t_{pLH}	Low-to-high propagation delay
t_r	Rise time
V_{BB}	Body biasing voltage
V_{dd}	Supply voltage
$V_{dd,MEP}$	Supply voltage at which the MEP occurs
$V_{dd,min}$	Minimal supply voltage
$V_{dd,nom}$	Nominal supply voltage
V_{ds}	Drain-source voltage of a transistor
V_{dsat}	Saturation drain voltage of a transistor
V_{FB}	Flatband voltage
V_{gs}	Gate-source voltage of a transistor
V_{IH}	Minimum high input voltage of an inverter
V_{IL}	Maximum low input voltage of an inverter
V_{in}	Input voltage
V_M	Switching threshold voltage of an inverter
V_{OH}	Minimum high output voltage of an inverter
V_{OL}	Maximum low output voltage of an inverter
V_{out}	Output voltage
V_{sb}	Source-bulk voltage of a transistor
V_{ss}	Ground voltage
V_T	Threshold voltage of a transistor
V_{T0}	Threshold voltage for $V_{sb} = 0$
V_{th}	Thermal voltage
W	Width of a transistor
α	Activity factor
γ	Body effect coefficient
Δ	Delay
ϵ_0	Permittivity of free space ($= 8.854 \cdot 10^{-14}$F/cm)
ϵ_i	Permittivity of a material i
η	DIBL coefficient
λ	Channel length modulation coefficient

μ	Mobility of the charge carriers
μ_x	Mean value of x
σ_x	Standard deviation of x
ϕ_F	Fermi potential

Chapter 1
Introduction

In today's society, portable electronic devices play a major role in everyday life. These portable systems demand an ever increasing energy-efficiency. Every year, a single system is expected to comprise more complexity while at the same time extending its autonomy. Many contradicting expectations are driving research toward more energy-efficient digital circuits. Since the evolution of the energy capacity of batteries only increases very slowly, energy-efficient circuits are key to reaching the ever increasing expectations of customers. Moreover, many new fields are emerging which have even more stringent requirements on energy-efficiency. Especially the medical world can greatly profit from today's evolutions. For instance, cheap sensor nodes and networks which can autonomously perform signal processing algorithms while demonstrating long lifetimes are becoming more and more feasible.

This sets the general context of this book, and will be further clarified in this chapter. How to achieve the goal of high energy-efficiency will be explained in the remainder of this book.

1.1 Supply Voltage Reduction: A Brief History

In 1965, Gordon Moore observed that the number of transistors on a chip had been increasing in an exponential manner [6]. He predicted that the transistor count on a chip would double every 18 months, at least for the next 10 years. Moore's law was born, and little did he know at the time that he had started a self-fulfilling prophecy. When looking back at CMOS technology scaling now, manufacturing companies have reduced transistor sizes with approximately $\sqrt{2}$ or 30 % with every technology node, to pack twice as many transistors in the same area [19].

N. Reynders, W. Dehaene, *Ultra-Low-Voltage Design of Energy-Efficient Digital Circuits*, Analog Circuits and Signal Processing,
DOI 10.1007/978-3-319-16136-5_1

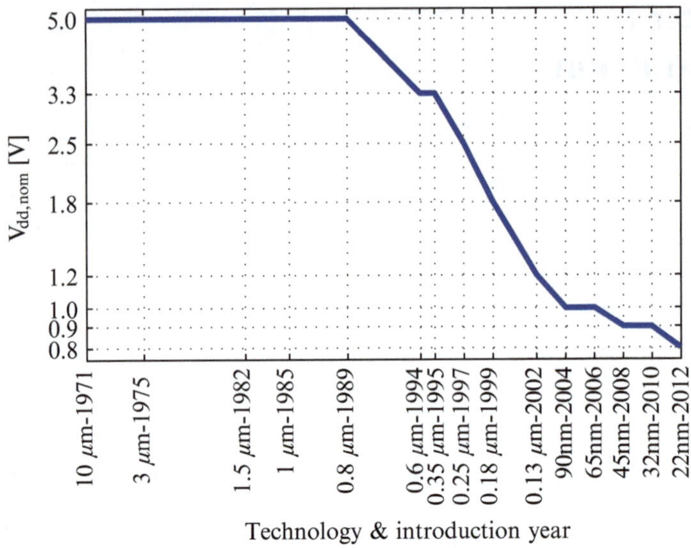

Fig. 1.1 Scaling of the nominal supply voltage of CMOS technology nodes over time [15, 20]

Ideally, by reducing transistor dimensions, delay reduces, circuits consume less power and energy, and manufacturing costs decrease. For digital systems, CMOS scaling has been driven by the increased performance that is obtained with every new technology node. Over time, the nominal supply voltage $V_{dd,nom}$ of those technology nodes has changed as well, as visualized in Fig. 1.1.

In general, scaling affects two independent variables, i.e. the geometric dimensions (e.g. of minimum devices) and the voltages (e.g. supply voltage V_{dd} and threshold voltage V_T). In the history of CMOS technologies, several scaling theories have dictated the evolution of $V_{dd,nom}$ in time.

Historically, Constant Voltage (CV) scaling was the norm. With this form of scaling, device dimensions shrink while the supply voltage remains constant. As can be seen in Fig. 1.1, a fixed supply voltage of 5 V has been employed until the end of the 1980s. The main advantage of using a constant voltage over different technology nodes was that circuits in newer technologies were still compatible with existing systems. This continuity in I/O voltages facilitated migrating designs to new technology nodes. With CV scaling, the delay—a driving force for scaling—does scale, but at the same time, power density increases considerably. Moreover, the electric fields in the devices increase as well, causing risk of device breakdown. Additionally, at a certain point, reducing the transistor size did not cause a decrease in delay anymore because of velocity saturation. More information on velocity saturation will be provided in Chap. 2. Therefore, CV scaling has not been pursued after the 0.8 μm node.

In 1974, Dennard proposed Constant Electric Field (CEF) scaling [3]. In this scaling theory, both voltage and geometric dimensions shrink proportionally to keep the electric fields constant. This also results in a power density which remains constant. CEF scaling avoids breakdown and thus ensures physical integrity. This scaling scenario is regarded as *ideal* scaling, as electric fields remain constant over different technology nodes.

This theory has not been adopted until the beginning of the 1990s, when CV scaling became unsustainable due to the problems mentioned above. Starting from the 0.6 μm technology node, the supply voltage has been scaled down almost every node. However, important to note is that CEF scaling has never been fully implemented. Although voltage reductions were realized, voltages scaled less than geometric dimensions. The reason for this is that several intrinsic device voltages are dependent on material characteristics and can therefore not be scaled [7]. Furthermore, scaling the threshold voltage results in a difficult trade-off between scaling fully proportionally and ensuring that the transistor can still be turned off.

Figure 1.1 shows another trend in nominal supply voltage scaling starting from the 90 nm technology node. As visible, $V_{\mathrm{dd,nom}}$ scaling slows down for advanced nanometer technologies. This is caused by the leakage increase which would become unacceptable if sustaining the prior scaling rate between 1989 and 2004. To accommodate for different customer needs, foundries began to offer two technology options starting from around the 130 nm technology node: a high-performance and a low-leakage option. The high-performance option tends to follow traditional scaling laws more than the low-leakage one [18]. When applicable, the $V_{\mathrm{dd,nom}}$ values depicted in Fig. 1.1 come from the high-performance processes. The question is now if this slowed down scaling rate for the nominal supply voltage can be continued for further technology nodes, or if unsustainable voltage scaling will put an end to CMOS technology scaling. Certainly, it introduces quite a challenge for technology scientists.

1.2 Reducing the Energy Consumption

CMOS technology scaling has traditionally been driven by cost and delay reductions. Historically, power and energy consumption decreased as well when scaling. However, the increase of static power consumption has become more and more pronounced in advanced nanometer technologies, eventually compromising the previous advantageous reduction of total power and energy consumption.

Additionally, more and more portable applications have emerged which are very often battery-powered. Since these batteries only possess a limited energy capacity, extending the lifetime of these battery-powered systems can only be achieved through consuming less energy. To realize more energy-efficient systems, a designer must gain insight into which mechanisms play a role in the power and energy consumption of a system.

1.2.1 Definitions

The total *power* consumption of a digital system consists of a dynamic and a static component:

$$P_{\text{tot}} = P_{\text{dyn}} + P_{\text{stat}} \tag{1.1}$$

The *dynamic* power dissipation is consumed when the system is actively switching and can generally be divided into short-circuit power and switching power:

$$P_{\text{dyn}} = P_{\text{short-circuit}} + P_{\text{switching}} \tag{1.2}$$

$P_{\text{short-circuit}}$ is the dissipation due to short-circuit currents which flow while there is a direct current path from the supply to the ground for a short period of time during switching [19]. Short-circuit power is strongly sensitive to the supply voltage. Therefore, its importance has reduced in time. It is now almost negligible for advanced nanometer technologies [7]. Since this book focuses on operating circuits at extremely low supply voltages, as will be explained later, $P_{\text{short-circuit}}$ is even more insignificant. This term will thus be ignored.

$P_{\text{switching}}$ is the power dissipation due to charging and discharging of load capacitances as logic gates switch:

$$P_{\text{switching}} = \alpha \cdot C \cdot V_{\text{dd}}^2 \cdot f_{\text{clk}} \tag{1.3}$$

with f_{clk} the clock frequency at which the system is working, V_{dd} the supply voltage, C the total load capacitance of the system, and α the activity factor. The activity factor is introduced because not all gates switch every clock cycle. Therefore, α is a measure of the average switching activity of the logic gates in a system. Its value lies between 0, when no gates ever switch, and 1, in which case all gates switch every clock cycle.

Static power consumption, on the other hand, is consumed even when a system is not switching. Through a transistor which is turned off still flows a small amount of leakage current, and this leakage current results in a static power dissipation. Hence, P_{stat} is defined by:

$$P_{\text{stat}} = I_{\text{leak}} \cdot V_{\text{dd}} \tag{1.4}$$

with I_{leak} as the total leakage current of the system. In advanced nanometer technologies, this leakage current increases due to the smaller feature size of the transistors. As mentioned above, P_{stat} is therefore increasingly important in these recent technologies, whereas in the early days of CMOS circuits, it could be ignored.

To summarize, the equation for the total power consumption of a digital system is:

$$P_{\text{tot}} = \alpha \cdot C \cdot V_{\text{dd}}^2 \cdot f_{\text{clk}} + I_{\text{leak}} \cdot V_{\text{dd}} \tag{1.5}$$

However, focusing on the power consumption of a system is not very relevant for most applications. Firstly, reducing the power consumption can be easily achieved by reducing the frequency and taking an infinite amount of time to finish an operation. Secondly, in battery-powered applications, it is not the power which counts, but the power consumed over an amount of time to perform an operation. This is called the *energy* consumption:

$$E_{\text{tot}} = P_{\text{tot}} \cdot t_{\text{clk}} \tag{1.6}$$

The total energy consumption of a digital system can then be calculated as follows:

$$\begin{aligned} E_{\text{tot}} &= E_{\text{dyn}} + E_{\text{stat}} \\ &= \alpha \cdot C \cdot V_{\text{dd}}^2 + I_{\text{leak}} \cdot V_{\text{dd}} \cdot t_{\text{clk}} \end{aligned} \tag{1.7}$$

1.2.2 Minimizing the Energy Consumption

Reducing the total energy consumption is of crucial importance for many applications. In order to design energy-efficient circuits, it is imperative to determine which component of energy is dominant for which type of circuit or system. As visible in (1.7), the energy consumption of digital circuits can be divided into two components, i.e. dynamic and static energy. The dynamic energy consists mostly of switching energy when load capacitances are charged or discharged. Static energy, on the other hand, is consumed as a result of all sources of leakage during a certain time period.

Both the dynamic and the static energy are dependent on the supply voltage. E_{dyn} is quadratically dependent on V_{dd} since none of the other parameters which determine the value of E_{dyn} are supply-dependent:

$$E_{\text{dyn}} \sim V_{\text{dd}}^2 \tag{1.8}$$

Hence, the dynamic energy reduces quadratically with decreasing V_{dd}. The static energy on the other hand is more complicated. The leakage current I_{leak} is independent of V_{dd} to the first order, not taking into account any secondary effects such as DIBL and channel length modulation (more information will be provided in Sect. 2.1.1). Therefore, static power (recall Eq. (1.4)) is linearly proportional to V_{dd}:

$$P_{\text{stat}} \sim V_{\text{dd}} \tag{1.9}$$

However, at the same time, the clock period t_{clk} is supply-dependent as well, as it is defined by:

$$t_{\text{clk}} \sim \frac{C \cdot V}{I} \tag{1.10}$$

The dependence of the current I on V_{dd} is determined by the operating region in which the transistors are functioning. In the weak inversion region, the current is exponentially dependent on the supply and therefore the following relations of t_{clk} and E_{stat} exist:

$$t_{\mathrm{clk}} \sim \frac{V_{\mathrm{dd}}}{\exp\left(V_{\mathrm{dd}}\right)} \tag{1.11}$$

$$E_{\mathrm{stat}} \sim \frac{V_{\mathrm{dd}}^2}{\exp\left(V_{\mathrm{dd}}\right)} \tag{1.12}$$

As opposed to E_{dyn}, E_{stat} thus increases with reducing V_{dd}. In the strong inversion region, on the other hand, the current is quadratically dependent on the supply:

$$t_{\mathrm{clk}} \sim \frac{V_{\mathrm{dd}}}{V_{\mathrm{dd}}^2} = \frac{1}{V_{\mathrm{dd}}} \tag{1.13}$$

Therefore, the static energy becomes independent of the supply voltage to the first order in the strong inversion region. To conclude, the graph of E_{stat} is flat at high supply voltages, not taking into account velocity saturation, but increases considerably in the weak inversion region. Hence, the supply voltage has opposing effects on the two components of the total energy, as E_{dyn} reduces quadratically with decreasing V_{dd} and E_{stat} increases exponentially in the weak inversion region. Therefore, there exists an optimal supply point $V_{\mathrm{dd,MEP}}$ at which the total energy is minimal. This is called the *Minimum-Energy Point* (MEP) [1, 17]. Since the drastic increase of the static energy occurs in the weak inversion region, the MEP is usually located in the sub- or near-threshold region.

At the MEP, static and dynamic energy are in balance. A circuit obtains its highest energy-efficiency at the MEP, since when it operates at this point, it consumes the least energy per operation. Important to realize is that the MEP is the least energy that an operation could consume if delay were unimportant [19]. Mathematically, the MEP occurs where the slopes of the static and dynamic energy are equal in magnitude and opposite in sign [18].

For a given system, the location of the MEP depends on the relative importance of dynamic versus static energy, which on its turn is strongly dependent on the activity of the system. A smaller activity factor leads to a reduction in dynamic energy, while static energy is unaffected by α. Figure 1.2 visualizes the evolution of the static, dynamic and total energy consumption with changing supply voltage for different activity factors: Fig. 1.2a has a high α and is therefore dominated by E_{dyn}, while Fig. 1.2a has a much lower α and is dominated by E_{stat}. The location of the MEP differs heavily between these two cases: it occurs at a much lower value of $V_{\mathrm{dd,MEP}}$ for the E_{dyn}-dominated case. An example of such a case are datapaths, while memories are for instance an excellent showcase for circuits which are dominated by static or leakage energy, since only few cells are accessed simultaneously.

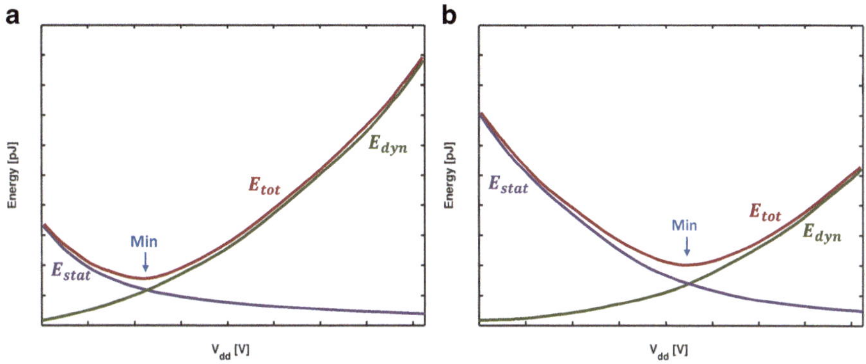

Fig. 1.2 Static, dynamic and total energy consumption as function of V_{dd}, for a system where (**a**) E_{dyn} dominates and (**b**) E_{stat} is dominant

Hence, extremely reducing the supply voltage is mostly interesting for circuits which are dominated by dynamic energy consumption. However, the feasible performance of such circuits decreases exponentially with V_{dd} (as will be explained thoroughly in Chap. 2). Therefore, possible applications for ultra-low-voltage operation are very energy-constrained systems which exhibit less severe performance constraints.

Decreasing the supply voltage and thus the dynamic energy only has a limited effect on leakage-dominated circuits, while the increased delay has a detrimental impact on the leakage energy. As a result, the MEP is located at a much higher $V_{dd,MEP}$.

To summarize, energy-efficient design is very dependent on the type of system. Depending on the application at hand, on the technology in which the system is fabricated, on its activity factor, etc., different measures should be taken to perform a given operation with minimal energy while still meeting the performance requirements. This book will focus on systems which are dominated by dynamic energy. In these systems, operating at ultra-low supply voltages can save orders of magnitude of energy dissipation.

1.2.3 Energy-Delay Product

As mentioned above, the MEP is an interesting metric as long as the delay at which this point occurs is still acceptable. However, applications often have delay constraints as well. When evaluating a system, it is therefore important to take this delay into account. A good metric to use for this is the *Energy-Delay Product* (EDP) [4], which combines a measure of performance and energy:

$$\mathrm{EDP} = E_{\mathrm{tot}} \cdot t_{\mathrm{clk}} \tag{1.14}$$

The EDP will be used as Figure Of Merit (FOM) in this book to compare obtained results of different designs. Since systems often trade energy for delay, or vice versa, only comparing on one of these specifications is insufficient.

1.3 Ultra-Low-Voltage Digital Design

Transistors are assumed to be ideal switches which conduct current for voltages above the threshold voltage and which are assumed to be turned off for voltages below V_T. However, in reality, this is a simplification of their actual behavior. The current does not abruptly cut off below V_T, but rather reduces exponentially with the supply voltage. Therefore, it is possible to operate circuits at supply voltages under or near the threshold voltage.

This has the advantage of possibly enabling drastic energy savings, as the MEP typically occurs at supply voltages much lower than the nominal supply, and is often located in the sub- or near-threshold region. As established in the previous section, high energy-efficiency can be achieved by operating a digital system at or around its MEP. The energy consumption of such systems can thus be drastically reduced by lowering their supply voltage. However, due to the decreased currents, it comes at the cost of a simultaneous increase in circuit delay. Furthermore, sub-threshold circuits suffer from an exponential sensitivity to variations. These and other challenges of ultra-low-voltage digital design will be addressed in detail in Chap. 2.

Already in 1972, Swanson et al. found that standard CMOS logic is functional for very low voltages [16]: by simulating the Voltage Transfer Characteristic (VTC) of an inverter and sweeping its supply voltage, the authors demonstrated that the inverter was able to provide gain for supply voltages as low as 100 mV. This paper is known as the first publication on digital circuits operating in the weak inversion region.

After this paper, there was some interest in operating transistors in the weak inversion region for analog design, but no attention has been given to sub-threshold digital design for a very long time. In the late '90s, interest for ultra-low-voltage digital circuits sparked again. For example, in 1999, Soeleman and Roy analyzed different logic families for sub-threshold operation [14]. In the twenty-first century, ultra-low-voltage digital research gained much more attention, which has been driven by the increasing amount of portable applications and by the increasing power problems due to scaling.

Figure 1.3 visualizes the increased popularity of ultra-low-voltage research since 1999. It shows the amount of publications per year which contain the words 'sub-threshold' or 'near-threshold' for CMOS technologies. As can be seen, early research uses 'sub-threshold' as catch phrase, while 'near-threshold' is recently becoming more and more the buzzword. This might seem confusing, but is entirely linked to the definition of the threshold voltage. The concept of V_T is mostly based on the previously discussed assumption of transistors as ideal

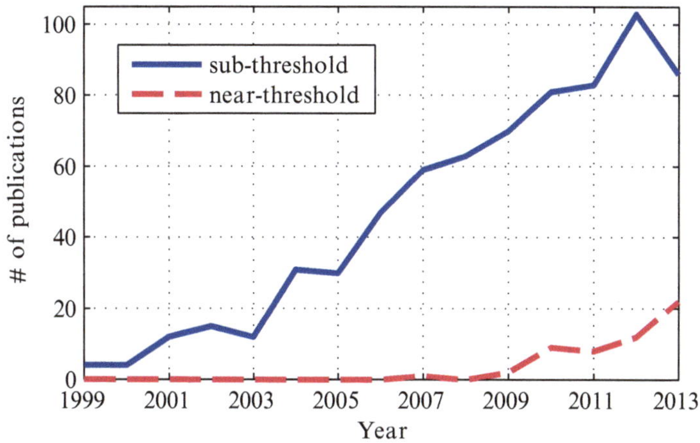

Fig. 1.3 Amount of publications per year that combine 'CMOS' with 'sub-threshold' or 'near-threshold' in their title or in the list of keywords on IEEE Xplore [5]

switches. In reality, the exact voltage at which the transition of operating regions occurs is difficult to define unambiguously, as will be discussed profoundly in Sect. 2.1.2. Therefore, naming a system as a sub- or near-threshold system has more to do with how to sell its functionality than with the literal interpretation of the name. In general, all these systems will be able to work at ultra-low supply voltages, and therefore in this book the term 'ultra-low-voltage' operation has been preferred.

However, operating digital systems at ultra-low supply voltages remains until now more or less an academic research field, since it has not yet been widely adopted by industry. There has been some interest in partially reducing the supply voltage of commercial applications with a few hundreds of mV below $V_{\mathrm{dd,nom}}$. When searching to reduce the energy consumption of a processor, these kind of measures can be taken without too much risk and with a lot of energy gain. However, understandably, companies are afraid to compromise the high yield of their systems by operating at extremely low supply voltages and hence suffering from an increased variability. To limit the risks, sub-threshold systems are often implemented with much overhead. This compromises on its turn the low-power and low-energy benefit of working at such low supply voltages. Therefore, it is often more beneficial to operate at a slightly higher supply voltage which does not require extreme measures to ensure robustness.

However, such large overheads can be avoided by designing *variation-resilient* circuits and systems. Hence, one of the goals of this work is to increase the industrial relevance of ultra-low-voltage circuits by guaranteeing a high yield through high variation-resilience of the system, and by being able to operate these circuits at speeds of $n \times 10\,\mathrm{MHz}$.

1.4 Applications

The energy consumption has become the critical parameter for a wide range of applications. This is where ultra-low-voltage digital circuits can play a large role, since those circuits are mostly adequate for applications with a very limited energy budget but less stringent speed performance requirements. Nonetheless, such applications often require moderate performance in the order of tens of MHz, and are not satisfied with only kHz-speed. This section will provide examples of such severely energy-constrained applications.

Firstly, there exists an increasing amount of *battery-powered* applications for which it is desirable to have an as prolonged lifetime as possible. Secondly, in some applications, the battery can be eliminated entirely by *harvesting energy* from the chip's environment. Obviously, a lower energy consumption to perform an operation is beneficial in this case as well.

Radio-Frequency Identification (RFID) tags fall into both categories: active RFID tags rely on their own power source, usually a battery, whereas passive RFID tags obtain power from an external source through energy harvesting [2]. In any case, both types of RFID tags are energy-constrained.

Many promising applications are situated in the medical world. Development of biomedical sensor nodes, such as wearable body patches for ECG monitoring of heart patients or measuring the blood glucose level of diabetics, is gaining more and more attention. Complete Body Area Networks (BAN) make it possible to record patient data in an efficient, comfortable manner. The growing market of implants requires an even more prolonged autonomy to reduce the amount of surgeries to a minimum. Examples are pacemakers, defibrillators and cochlear implants. The sound processor in these cochlear implants is a Digital Signal Processor (DSP) with algorithms for noise suppression and speech analysis and thus has quite high demands for computing power.

Many other sensor networks are being developed at the moment, for instance in the automotive industry, for monitoring systems, in smart warehouses, etc. In future concepts such as ubiquitous computing, the Internet of Things and ambient intelligence, an inconceivable amount of chips are foreseen which are all expected to be energy-autonomous or consume ultra-low amounts of energy. The realization of such futuristic concepts relies on an ever growing complexity at an ever reducing energy consumption. As a result, energy-efficiency is the key to the future. Ultra-low-voltage circuit design is a very promising research field to fulfill these future requirements.

1.5 Current State-of-the-Art in Literature

This section will provide an overview of the current state-of-the-art in literature on ultra-low-voltage digital circuit design in CMOS technologies. A subset of papers will be discussed. The criterion for their selection is if they have provided

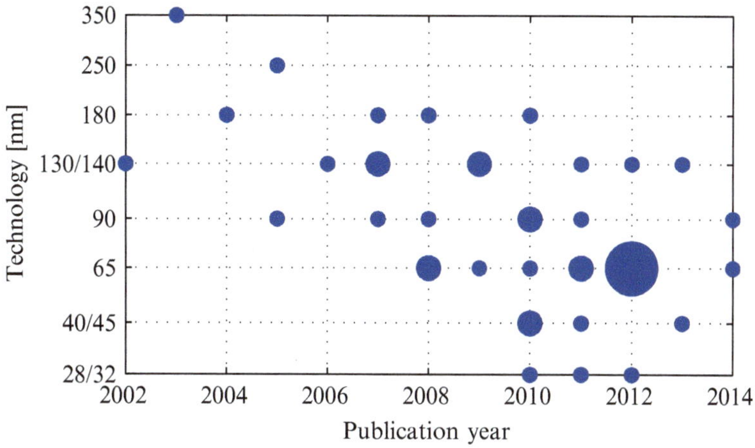

Fig. 1.4 CMOS technologies of all state-of-the-art publications with ultra-low-voltage measurement results as function of their publication year. The size of the circles is proportional to the population size

measurement results of a substantial digital circuit which functions at supply voltages below 500 mV. Therefore, the implementations of the papers range from adders and multipliers to full DSPs and microcontrollers. An extensive table containing the paper details and the measured operating points can be found in Appendix A.

As with all digital design, this field follows the current scaling trends. This can be seen in Fig. 1.4, which plots the CMOS technologies of the publications as function of the year in which they are published. The size of the circles is proportional to the amount of publications in that technology in that year. A general downwards scaling trend is clearly visible.

The minimal supply voltage $V_{\mathrm{dd,min}}$ at which these designs are reported to be functional is given as function of CMOS technology node in Fig. 1.5a. The figure also indicates whether a design used body biasing or not. Body biasing is a form of threshold voltage manipulation. It is sometimes used in ultra-low-voltage design to cope with inter-die variations. More information on body biasing will be provided further on in Sects. 2.1.2.2 and 3.1.1.6. It is apparent that body biasing has been used in the older technology nodes, and is thereafter only occasionally employed. What is important to notice, is that no clear trend is visible between the designs which do use body biasing and the ones which do not. This has to do with the fact that the $V_{\mathrm{dd,min}}$ of ultra-low-voltage systems is mostly restricted due to individual variations on each transistor than to global variations across all transistors.

Note that body biasing might become more useful in new technologies, such as e.g. fully depleted Silicon-On-Insulator (SOI), but this is out of the scope of this book.

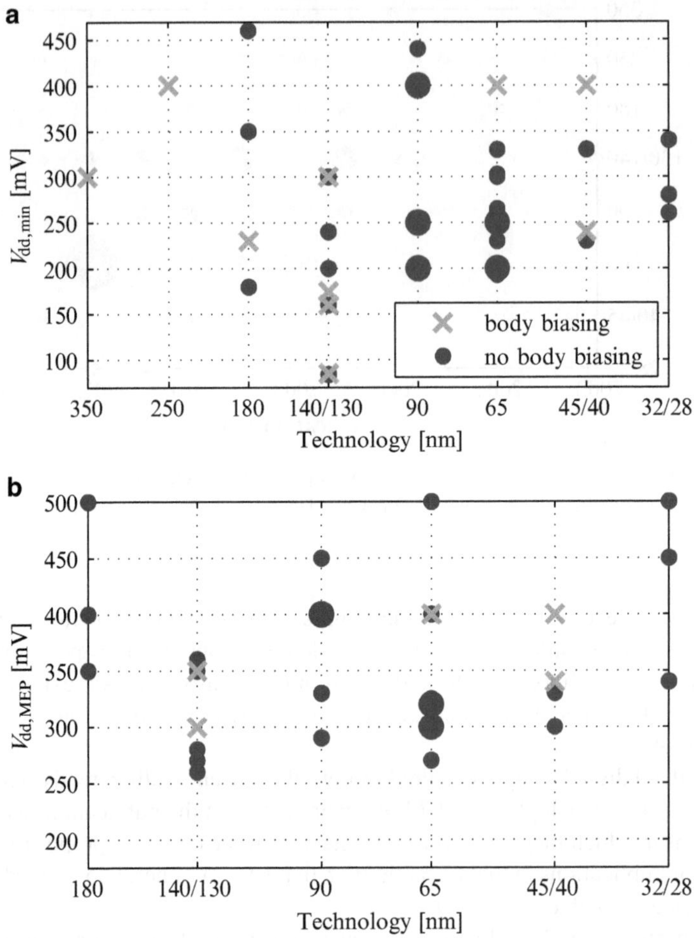

Fig. 1.5 Key voltages as function of CMOS technology node, with marker size proportional to population size: (**a**) minimal functional supply voltage $V_{dd,min}$ and (**b**) supply voltage at which the MEP occurs $V_{dd,MEP}$. A division is made between designs which use body biasing and those which do not

The minimal supply voltage at which a design is functional is an interesting metric, as it gives an idea of how variation-resilient a design is. Generally speaking, the lower the supply voltage, the more variability becomes important relatively. Therefore, the lower $V_{dd,min}$ a design achieves, the more it can cope with these variations. Of course, this is technology-dependent, so simply comparing different achieved $V_{dd,min}$ values of designs in different technologies is not fair. Nonetheless, $V_{dd,min}$ does provide a certain measure of variation-resilience.

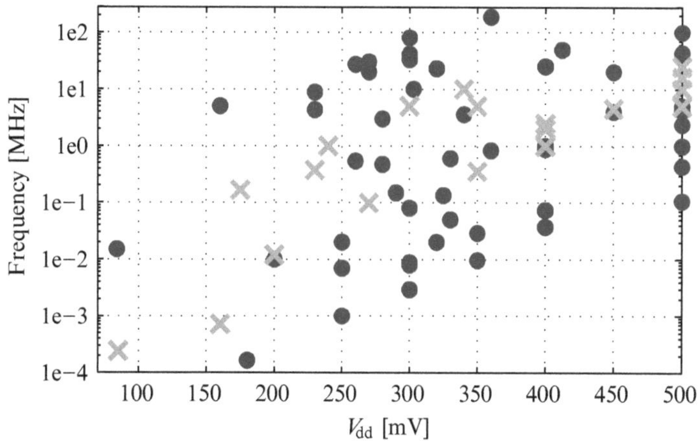

Fig. 1.6 Frequency as function of V_{dd} for all provided measurement points of the publications of Appendix A, for designs in all CMOS technology nodes

From Fig. 1.5a, it can be seen that apart from two designs in the 130/140 nm node which are operating right below 100 mV, all designs have a $V_{dd,min}$ of more than 150 mV, and most of them above 200 mV. However, no clear trend is visible between $V_{dd,min}$ and technology scaling.

This is the case for the supply voltage at which the MEP occurs as well, as visible in Fig. 1.5b. Since not all publications of Appendix A provided the MEP data, less data points are visible in this graph. The importance of the MEP has been accepted starting from the 180 nm technology node, as can be seen. Again, no pattern is visible for technology or for body biasing.

Body biasing is often used to increase performance at the cost of increased leakage. Do the designs with body biasing then perform better in speed? The following graph, Fig. 1.6, provides an overview of all frequency measurement points which were given in the publications of Appendix A. Note that this graph contains designs in all the different CMOS technology nodes which were shown in Fig. 1.4. Hence, this figure is mostly useful to show the speed trends in ultra-low-voltage designs, and less to compare exact measurement points. Nonetheless, one can see that a large amount of the published designs only achieved kHz-speed, even at higher supply voltages. Furthermore, the designs with body biasing do not show a significant improvement in speed compared to the other designs, making the benefits of the use of body biasing for ultra-low-voltage digital design in CMOS technologies questionable.

To make ultra-low-voltage digital circuit design more attractive for industry, higher operating frequencies are required. The area of interest of this book is therefore the upper left triangle of Fig. 1.6, where speeds well within the MHz-range are achieved at ultra-low supply voltages, provided that the energy consumption is still maintained low.

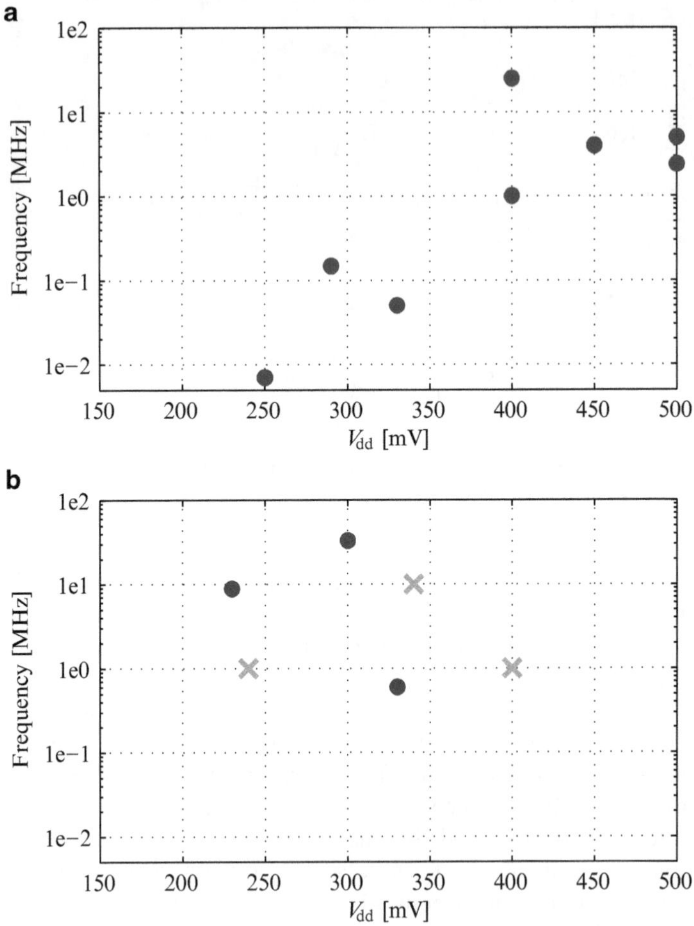

Fig. 1.7 Frequency as function of V_{dd} for all provided measurement points of the publications of Appendix A in specific technology nodes: (**a**) 90 nm CMOS and (**b**) 40 nm CMOS

This work focuses on ultra-low-voltage digital design in bulk CMOS technologies. Prototypes in two different CMOS technologies will be presented: the 90 nm node and the 40 nm node. In both cases, the designs are performed in the high-performance process of the technologies.

Figure 1.7a and b visualize the subsets of measurement points of Fig. 1.6 in both technologies at hand of this book. In the comparison of Chap. 7, the results of the prototypes of this book shall be compared with the current state-of-the-art in literature presented here.

While it is meaningful to compare the operating frequency of different digital circuits, this is not the case for their energy consumption, since these systems perform different operations. Therefore, no overall graph which shows the energy consumption is included here. However, whenever possible throughout this work, the energy consumption will be compared when it consists of similar designs and hence produces meaningful comparisons.

1.6 Outline of This Work

To conclude, this book aims to develop circuit and architectural techniques to design ultra-low-voltage digital circuits with high energy-efficiency in bulk CMOS technologies. Targets are signal processing applications, since DSPs are often necessary to increase the processing power of portable sensor applications. This processing power can then be employed to perform data processing and compression locally in the sensor node. To allow a more widespread use of ultra-low-voltage digital circuits, it is essential that operating frequencies well beyond 10 MHz are achieved. Additionally, high variation-resilience is crucial to cope with the increased variability of circuits operating at extremely low supply voltages and hence to obtain high yield.

Figure 1.8 provides a graphical overview of the relationship between the different chapters of this work. The division is closely related to the design abstraction levels of digital systems. To be able to design ultra-low-voltage circuits which are

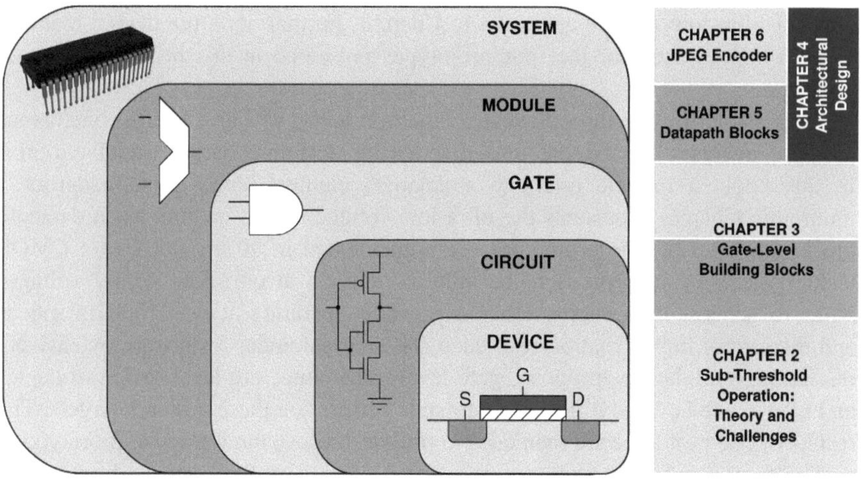

Fig. 1.8 Graphical overview of this work: the chapters follow the general design abstraction levels of digital systems

not only energy-efficient, but also achieve an acceptably high performance, while guaranteeing variation-resilience, design choices must be made on all abstraction levels. Therefore, a bottom-up approach has been used in this book.

To start, it is important to gain insight into sub-threshold behavior of transistors in CMOS technologies. Therefore, Chap. 2 discusses the different operating regions of a CMOS transistor. It examines the threshold voltage definition of the device, as well as the parameters which influence V_T. While Chap. 1 has mostly highlighted the advantages of operating circuits at ultra-low supply voltages, Chap. 2 provides an overview of the different challenges which are introduced by operating in this region. In this book, prototypes have been designed and fabricated in two different CMOS technologies. Hence, the impact of scaling is studied on device-level in this chapter as well. Finally, the type of CMOS technologies and transistors which is preferred and employed in this work is revealed.

After exploring the consequences of ultra-low-voltage operation on device-level, Chap. 3 investigates the functionality of various circuit topologies at ultra-low supply voltages. These different logic families are extensively compared on numerous circuit characteristics. An in-depth analysis of their performance to these characteristics leads to the presentation of the circuit topologies which will be preferred in this work [9]. This chapter not only introduces preferred implementation of logic gates, but of memory elements as well. As such, all gate-level building blocks which will be employed in the prototypes of this work are presented.

Before proceeding to the subsequent abstraction levels, an architectural framework for these module- and system-level digital circuits needs to be established. Therefore, the benefits and drawbacks of different architectural options are examined for ultra-low-voltage operation in Chap. 4. Furthermore, the design methodology which is used for the four prototypes presented in this book is discussed profoundly.

After exploring the three lowest abstraction levels of Fig. 1.8, the conclusions of these analyses can be put into practice by designing larger digital circuits. In this book, a module refers to a datapath element, such as an adder or a multiplier. Chapter 5 presents the ultra-low-voltage design of three such datapath blocks [8, 10, 12]. The prototypes are implemented in 90 nm and 40 nm CMOS technologies. Their target is to be able to function at ultra-low supply voltages so as to achieve high energy-efficiency, while operating at $n \times 10\,\text{MHz}$ speeds and displaying high variation-resilience. The measurement results are extensively discussed to validate the proposed gate-level and architecture-level design strategies and to evaluate how the different prototypes perform on these research targets. The results of one prototype are then used to further improve the following prototype.

The fourth and final prototype is situated on system-level and is discussed in Chap. 6. It consists of a full JPEG encoder designed and fabricated in a 40 nm CMOS technology [11, 13]. Its purpose is to validate that the proposed design strategy of this book is generally applicable in any large and complex ultra-low-voltage DSP design. The different design efforts to accomplish a functional DSP system at ultra-low supply voltages are extensively discussed. The measurement results are again examined profoundly and compared to the state-of-the-art.

Chapter 7 concludes this work by returning to the current state-of-the-art in literature which has been presented in Chap. 1 and by evaluating how the prototypes of this work perform when compared to these designs. A general conclusion is provided and the main contributions of this work are listed. Furthermore, suggestions for future work and improvements are given.

References

1. Calhoun B, Chandrakasan A (2004) Characterizing and modeling minimum energy operation for subthreshold circuits. In: Proceedings of the ACM/IEEE international symposium on low power electronics and design (ISLPED), pp 90–95. DOI: 10.1109/LPE.2004.1349316
2. Dehaene W, Gielen G, Steyaert M, Danneels H, Desmedt V, De Roover C, Li Z, Verhelst M, Van Helleputte N, Radiom S, Walravens C, Pleysier L (2009) RFID, where are they? In: Proceedings of the IEEE European solid-state circuits conference (ESSCIRC), pp 36–43. DOI: 10.1109/ESSCIRC.2009.5325928
3. Dennard RH, Gaensslen F, Yu HN, Rideout L, Bassous E, Leblanc AR (1974) Design of ion-implanted MOSFET's with very small physical dimensions. IEEE J Solid-State Circuits SC-9(5):256–268
4. Gonzalez R, Gordon B, Horowitz M (1997) Supply and threshold voltage scaling for low power CMOS. IEEE J Solid-State Circuits 32(8):1210–1216. DOI: 10.1109/4.604077
5. IEEE IEEE Xplore digital library. URL http://ieeexplore.ieee.org
6. Moore GE (1965) Cramming more components onto integrated circuits. Electronics 38(8):114–117
7. Rabaey J, Chandrakasan A, Nikolic B (2003) Digital integrated circuits: a design perspective, 2nd edn. Prentice Hall, Upper Saddle River, New Jersey
8. Reynders N, Dehaene W (2011) A 190mV supply, 10MHz, 90nm CMOS, pipelined subthreshold adder using variation-resilient circuit techniques. In: Proceedings of the IEEE Asian solid-state circuits conference (A-SSCC), pp 113–116, DOI: 10.1109/ASSCC.2011.6123617
9. Reynders N, Dehaene W (2012) Variation-resilient building blocks for ultra-low-energy subthreshold design. IEEE Trans Circuits Syst–Part II: Express Briefs 59(12):898–902. DOI: 10.1109/TCSII.2012.2231022
10. Reynders N, Dehaene W (2012) Variation-resilient sub-threshold circuit solutions for ultra-low-power digital signal processors with 10MHz clock frequency. In: Proceedings of the IEEE European solid-state circuits conference (ESSCIRC), pp 474–477. DOI: 10.1109/ESSCIRC.2012.6341358
11. Reynders N, Dehaene W (2014) A 210mV 5MHz variation-resilient near-threshold JPEG encoder in 40nm CMOS. In: Proceedings of the IEEE international solid-state circuits conference (ISSCC), pp 456–457
12. Reynders N, Dehaene W (2015) On the effect of technology scaling on variation-resilient subthreshold circuits. Elsevier Solid-State Electron 103:19–29
13. Reynders N, Rooseleer B, Dehaene W (2014) Energy-efficient logic and SRAM design: A case study. In: Proceedings of the IEEE faible tension faible consommation conference (FTFC), pp 1–4. DOI: 10.1109/FTFC.2014.6828616
14. Soeleman H, Roy K (1999) Ultra-low power digital subthreshold logic circuits. In: Proceedings of the ACM/IEEE international symposium on low power electronics and design (ISLPED), pp 94–96
15. Stanford University VLSI Research Group CPU database. URL http://cpudb.stanford.edu/
16. Swanson R, Meindl J (1972) Ion-implanted complementary MOS transistors in low-voltage circuits. IEEE J Solid-State Circuits 7(2):146–153, DOI: 10.1109/JSSC.1972.1050260

17. Wang A, Chandrakasan A, Kosonocky S (2002) Optimal supply and threshold scaling for subthreshold CMOS circuits. In: Proceedings of the IEEE computer society annual symposium on VLSI (ISVLSI), pp 5–9. DOI: 10.1109/ISVLSI.2002.1016866
18. Wang A, Calhoun B, Chandrakasan A (2006) Sub-threshold design for ultra low-power systems. Springer, New York
19. Weste N, Harris D (2011) CMOS VLSI design: a circuits and systems perspective, 4th edn. Addison-Wesley, New York
20. Wikipedia Semiconductor device fabrication. URL http://en.wikipedia.org/wiki/Semiconductor_device_fabrication

Chapter 2
Sub-Threshold Operation: Theory and Challenges

In order to design circuits operating at ultra-low supply voltages, an understanding about the sub-threshold behavior of CMOS transistors must first be obtained. Therefore, this chapter discusses the fundamentals of sub-threshold operation, by looking at the general principles of transistor theory and by briefly giving some adequate background of the device physics in Sect. 2.1. The different operating regions of a CMOS transistor are examined, as well as the definition of the threshold voltage and the parameters by which it is influenced.

The main advantages of sub-threshold circuits were already discussed in Chap. 1. Section 2.2 of this chapter provides an overview of the different challenges which are introduced by operating a circuit in the sub-threshold or weak inversion region. Insight in these circuit-level challenges is essential to efficiently and successfully design ultra-low-voltage systems.

Subsequently, the impact of CMOS technology scaling on circuits operating in the ultra-low-voltage region is studied. Section 2.3 aims to provide an answer to the benefits and disadvantages of scaling on such implementations [29]. First, an equation to determine the minimum feasible supply voltage for digital circuits is derived. Out of this equation, a theoretical minimum as well as a practical minimum supply for a specific technology can be calculated. Second, scaling analysis focuses on the two CMOS technologies at hand of this book. Furthermore, the (in)accuracy of weak inversion transistor models will also be explored.

To conclude, Sect. 2.4 explains what the difference is between the various transistor types offered by modern CMOS technologies and which type of transistors is used throughout the prototypes presented in this work. Finally, Sect. 2.5 concludes this chapter.

Simulation results in this chapter are obtained with the 90 nm CMOS technology at hand, unless stated otherwise.

N. Reynders, W. Dehaene, *Ultra-Low-Voltage Design of Energy-Efficient Digital Circuits*, Analog Circuits and Signal Processing,
DOI 10.1007/978-3-319-16136-5_2

2.1 Transistor Operation

Figure 2.1 shows the cross-section of a typical nMOS transistor when no voltages are applied. It also shows the four terminals through which voltages can be applied, i.e. the source (S), drain (D), gate (G) and bulk (B) terminals. When a small positive gate-source voltage V_{gs} is applied, the holes in the region of the substrate below the oxide (also called the surface) are repelled from the gate, leaving behind positively charged immobile atoms. The resulting *depletion region* can be seen in Fig. 2.2a.

The drain and source form two *np* junctions with the bulk. When the potential difference V_{ds} between the drain and the source is positive, the reverse bias across the *np* junction of the drain is larger than the one across the source-bulk junction, thereby resulting in a deeper depletion region at the drain (see Fig. 2.2a) [34].

When V_{gs} reaches a critical value, i.e. V_{gs} exceeds the threshold voltage V_T, the surface becomes attractive to electrons from the n^+ regions and the depletion region stops growing. Free electrons flow from the source to the drain and form an *inversion layer* or *channel* under the gate oxide. For $V_{ds} \leq V_{gs} - V_T$, the voltage at any point in the channel is larger than the threshold voltage and the drain-source current I_{ds} increases linearly with V_{ds}. This is called the *linear* region of the transistor and is

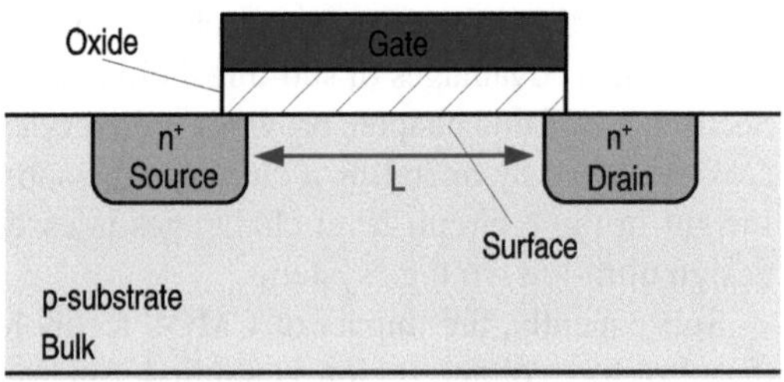

Fig. 2.1 Cross-section of a typical nMOS transistor

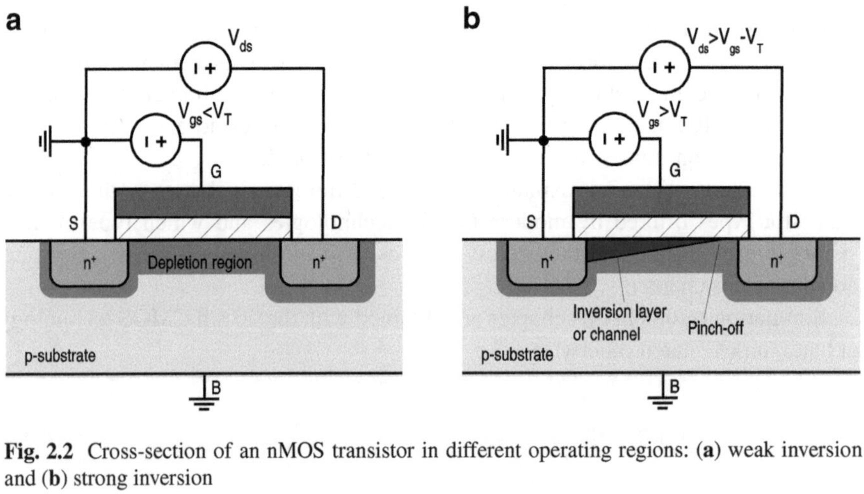

Fig. 2.2 Cross-section of an nMOS transistor in different operating regions: (**a**) weak inversion and (**b**) strong inversion

not an interest of this book. However, for $V_{ds} > V_{gs} - V_T$, the assumption that the channel voltage is larger than the threshold voltage all along the channel ceases to hold. Under those circumstances, the transistor is in the *saturation* region. The point where the conducting channel disappears is the *pinch-off* point [26]. This can be seen in Fig. 2.2b.

From now on, this chapter will concentrate on the dependence of the drain-source current I_{ds} on the gate-source voltage V_{gs}.

2.1.1 Different Operating Regions

The normalized drain-source current I_{ds} of a typical MOS transistor as function of the gate-source voltage V_{gs} is shown in Fig. 2.3. Three different operating regions of the transistor can be distinguished: exponential, quadratic and linear behavior of the current. These regions are respectively called the weak inversion region, the strong inversion region and the velocity saturation region, and will now be discussed.

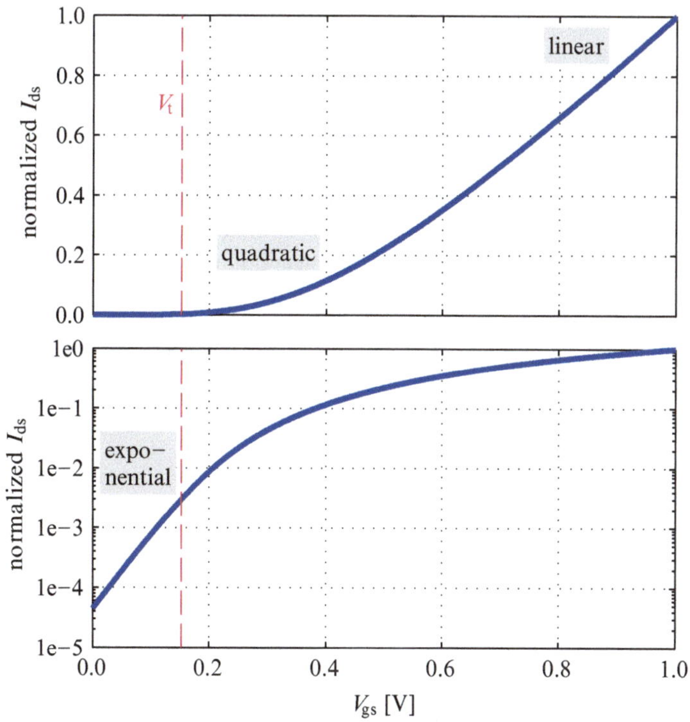

Fig. 2.3 Normalized current I_{ds} as function of V_{gs} of a typical minimal MOS transistor ($V_{ds} = 1$ V) in nominal operation

2.1.1.1 Strong Inversion

When V_{gs} is larger than V_T, a channel of charge carriers is formed under the gate oxide between the source and the drain. Since $V_{ds} > V_{gs} - V_T$, pinch-off occurs. The transistor is then said to operate in the *strong inversion* region, as shown in Fig. 2.2b. The conduction is dominated by the drift current, which depends on the applied electric field that exerts forces on the charged carriers (thus the potential difference V_{ds}) and also depends on the mobility of the carriers.

The equation that describes the quadratic relationship of the current with respect to V_{gs} (Fig. 2.3) in the strong inversion region is [26, 27, 38]:

$$I_{ds,si} = \frac{1}{2} \cdot \mu \cdot C_{ox} \cdot \frac{W}{L} \cdot \left(V_{gs} - V_T\right)^2 \cdot (1 + \lambda \cdot V_{ds}) \tag{2.1}$$

with W and L the width and length of the transistor channel, V_T the threshold voltage of the transistor and C_{ox} the gate oxide capacitance per unit area, defined by:

$$C_{ox} = \frac{\epsilon_{ox}}{t_{ox}} \tag{2.2}$$

where t_{ox} is the gate oxide thickness and ϵ_{ox} the permittivity of the oxide, given by:

$$\epsilon_{ox} = K_{ox} \cdot \epsilon_0 \tag{2.3}$$

In (2.3), ϵ_0 is the permittivity of free space and K_{ox} is the dielectric constant of the gate oxide.

In (2.1), μ is the mobility of the electrons for an nMOS transistor and of the holes for a pMOS transistor. The mobility of holes in silicon is typically 2–3 times lower than the mobility of electrons, meaning that pMOS transistors provide less current than nMOS transistors of equal size. The last factor of (2.1) is added to include *channel length modulation*, with λ as the channel length modulation coefficient. This is the phenomenon in which a higher V_{ds} increases the size of the depletion region around the drain and thus effectively shortens the length of the conductive channel. Channel length modulation is typically more pronounced in shorter transistors.

2.1.1.2 Velocity Saturation

The state of *velocity saturation* is a typical behavior of short-channel transistors. When the electric field along the channel reaches a critical value, the velocity of the carriers tends to saturate due to scattering effects because of the collisions suffered by the carriers [26]. The point at which the electric field in the channel ultimately reaches its critical value is called the saturation drain voltage V_{dsat}.

As a consequence, from that point I_{ds} increases linearly with V_{gs} instead of quadratically (see Fig. 2.3):

$$I_{ds,vs} = \mu \cdot C_{ox} \cdot \frac{W}{L} \cdot \left(\left(V_{gs} - V_T \right) \cdot V_{dsat} - \frac{V_{dsat}^2}{2} \right) \cdot (1 + \lambda \cdot V_{ds}) \qquad (2.4)$$

2.1.1.3 Weak Inversion

The region of interest of this work, the *weak inversion* region, is also called the *sub-threshold* region, as a transistor is already partially conducting for voltages below the threshold voltage. Unlike the strong inversion region in which the drift current dominates, the sub-threshold conduction is dominated by diffusion current [30]. This diffusion current occurs when there are concentration gradients and particles are not distributed uniformly over space. Diffusion is not due to electric fields, and therefore I_{ds} will not depend on V_{ds} in first order. Figure 2.2a shows an nMOS transistor in the weak inversion region.

In this region ($V_{gs} \leq V_T$), the current I_{ds} is exponentially dependent on V_{gs}, as can be seen in Fig. 2.3. The equation that describes the weak inversion current is [30]:

$$I_{ds,wi} = I_0 \cdot \exp\left(\frac{V_{gs} - V_T}{n \cdot V_{th}} \right) \cdot \left(1 - \exp\left(\frac{-V_{ds}}{V_{th}} \right) \right) \cdot (1 + \lambda \cdot V_{ds}) \qquad (2.5)$$

where V_{th} is the thermal voltage, defined by:

$$V_{th} = \frac{kT}{q} \qquad (2.6)$$

with k as the Boltzmann constant, T the absolute temperature and q the electric charge of an electron. V_{th} is 25.85 mV at room temperature (300 K). In (2.5), n is a process-dependent parameter and I_0 is the current when $V_{gs} = V_T$. I_0 is dependent on process and device geometry [30, 37, 38]:

$$I_0 = \mu \cdot C_{ox} \cdot \frac{W}{L} \cdot (n - 1) \cdot V_{th}^2 \qquad (2.7)$$

The third factor of (2.5) indicates that $I_{ds,wi}$ is 0 if $V_{ds} = 0$, and only has an influence when V_{ds} is smaller than a few multiples of V_{th} because the factor reaches its full value of 1 from then on. In the remainder of this work, this third factor, which incorporates the current roll-off, will be omitted since it only has an effect when V_{ds} drops to within a few times V_{th}.

Because (2.5) is then only linearly dependent on V_{ds} (through channel length modulation) as opposed to the exponential dependence on V_{gs}, the first-order approximation of the weak inversion current which will be used in the remainder of this work also omits the fourth factor, thereby leading to:

$$I_{\text{ds},wi} \approx I_0 \cdot \exp\left(\frac{V_{\text{gs}} - V_{\text{T}}}{n \cdot V_{\text{th}}}\right) \tag{2.8}$$

2.1.2 *Threshold Voltage*

The *threshold voltage* V_{T} is the value of V_{gs} where the transistor changes from the weak inversion to the strong inversion region. Specifically, it is the point where the drift current starts to dominate over the diffusion current. So far, V_{T} was treated as a constant value. However, the threshold voltage is influenced by various parameters, which will be discussed in this section.

2.1.2.1 Definition

The equation that defines V_{T} is the following: [9, 34]

$$V_{\text{T}} = V_{\text{T0}} + \underbrace{\gamma \cdot \left(\sqrt{\phi_0 + V_{\text{sb}}} - \sqrt{\phi_0}\right)}_{\text{body effect}} - \underbrace{\eta \cdot V_{\text{ds}}}_{\text{DIBL}} - \underbrace{\Delta V_{\text{T}}}_{\text{SCE}} \tag{2.9}$$

where V_{T0} is the threshold voltage for source-bulk voltage V_{sb} equal to 0, defined by:

$$V_{\text{T0}} = V_{\text{FB}} + \phi_0 + \gamma \cdot \sqrt{\phi_0} \tag{2.10}$$

In (2.10), the flatband voltage V_{FB} is the difference between the work functions of the polysilicon gate and the silicon substrate [27]. The parameter ϕ_0 in (2.9) and (2.10) denotes the onset of strong inversion and is often taken equal to $2\phi_F$ (e.g. in [21, 26, 31, 38]), but this is not justifiable because it is larger than $2\phi_F$ in practice. The interested reader is referred to Sect. 2.6.2 of [34] for a more elaborate discussion on this subject. To summarize, a more appropriate value for ϕ_0 is:

$$\phi_0 = 2\phi_F + \Delta\phi \tag{2.11}$$

with $\Delta\phi$ as a constant term and ϕ_F as the so-called Fermi potential which is a quantity that characterizes a semiconductor material at a given temperature, defined by:

$$\phi_F = V_{\text{th}} \cdot \ln\left(\frac{N_{\text{A}}}{n_i}\right) \tag{2.12}$$

with N_A as the doping concentration of the substrate and n_i as the intrinsic carrier concentration. From (2.12), it is obvious that V_T is strongly dependent on the temperature through V_{th} in ϕ_F.

The temperature and other parameters which influence the threshold voltage will now be discussed in detail.

2.1.2.2 Body Effect

A first major influence on V_T is the *body effect* through the source-bulk voltage V_{sb}, as indicated in (2.9). When a positive voltage V_{sb} is applied between the source and bulk of an nMOS transistor, it increases the amount of charge required to invert the channel, hence, it increases the threshold voltage [38]. This is called Reverse Body Biasing (RBB). On the other hand, when a negative voltage V_{sb} is applied, V_T decreases. This is called Forward Body Biasing (FBB). Figure 2.4 shows the normalized threshold voltage of a typical nMOS as function of V_{sb}.

The value of the body effect coefficient γ determines how much effect changes in V_{sb} have. It is typically in the range of 0.4 to 1.0 $V^{1/2}$ [38] and is defined by:

$$\gamma = \frac{\sqrt{2 \cdot q \cdot \epsilon_{Si} \cdot N_A}}{C_{ox}} \tag{2.13}$$

with ϵ_{Si} as the permittivity of silicon which can be calculated in (2.3) by inserting $K_{Si} \approx 11.9$ [34].

The body effect is sometimes used as a means to perform threshold voltage manipulation. Different body biasing techniques are discussed in Sect. 3.1.1.6.

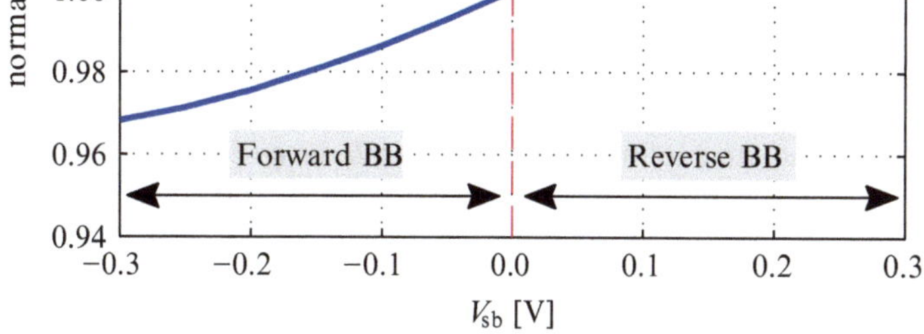

Fig. 2.4 Simulated normalized threshold voltage V_T as function of V_{sb} of a typical minimal nMOS transistor

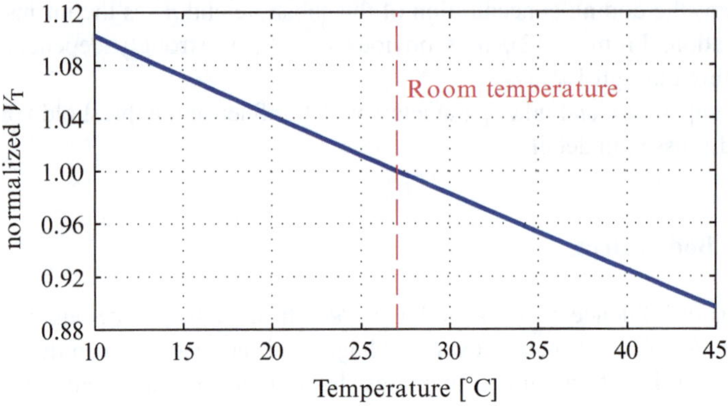

Fig. 2.5 Simulated normalized threshold voltage V_T as function of temperature of a typical minimal nMOS transistor

An important side note is that the impact of body biasing reduces for short-channel devices. The contribution of the depletion regions of the source and drain to the edges of the depletion region becomes relatively more important for shorter channel lengths. Therefore, the dependence of the threshold voltage on the body bias becomes weaker as channel length becomes shorter, because the body bias has less control of the depletion region due to the edge effects. To conclude, the influence of body biasing decreases with scaling.

2.1.2.3 Temperature

The threshold voltage is strongly dependent on the temperature due to the inclusion of the thermal voltage V_{th} in ϕ_0, thereby influencing both V_{T0} and the body effect in (2.9). As temperature is raised, the threshold voltage decreases almost linearly, as shown in Fig. 2.5. The temperature sensitivity of V_T is typically in the order of 0.5 to 3.0 mV/°C [30, 34]. For the nMOS transistor of a 90 nm technology in Fig. 2.5, it is about 0.7 mV/°C.

2.1.2.4 Drain-Induced Barrier Lowering

A third phenomenon that influences V_T is *Drain-Induced Barrier Lowering* (DIBL). There is a potential barrier at the surface between the source and the drain which prevents electrons from flowing from the source to the drain in off-conditions. For a long-channel device, the height of this barrier is mainly controlled by V_{gs} as the source and drain depletion regions are separated far enough so that V_{ds} has no effect. The threshold voltage of a long-channel device is thus independent of the channel length and drain voltage. However, for a short-channel device, the

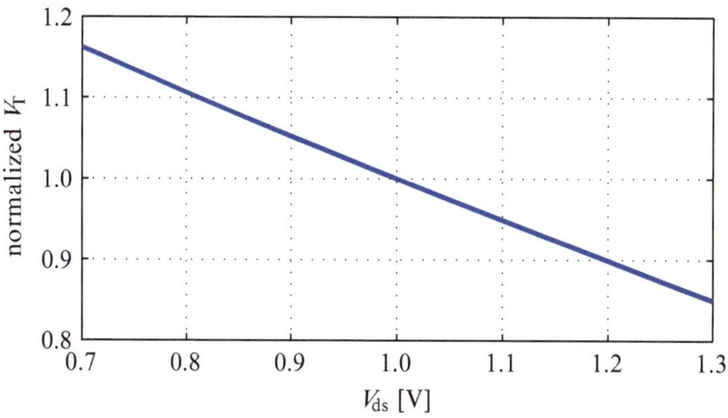

Fig. 2.6 Simulated normalized threshold voltage V_T as function of V_{ds} of a typical minimal nMOS transistor to illustrate the effect of DIBL

source and drain depletion regions approach each other enough so that they start to interact near the channel surface resulting in a lower source potential barrier [30]. Increasing V_{ds} results in a deeper depletion region at the drain (see Fig. 2.2a) and in a higher proximity to the depletion region at the source, thereby lowering the barrier height. To summarize, DIBL occurs in short-channel devices and the effect increases with V_{ds}. The threshold voltage thus reduces with an increasing V_{ds}, as shown in Fig. 2.6.

The term that introduces DIBL in (2.9) includes V_{ds} and the DIBL coefficient η, which is typically in the order of 100 mV/V [38].

2.1.2.5 Short-Channel Effect

In the traditional derivation of the threshold voltage, it is assumed that V_T is independent of the channel length and drain voltage (the dependence on the drain voltage due to DIBL has been addressed above). The depletion region is assumed to be controlled solely by the applied gate voltage and the depletion regions around the source and drain are ignored. However, as the channel length becomes shorter and shorter, this assumption ceases to hold. The depletion regions of the source and drain become relatively more important with shrinking channel lengths, introducing edge effects which cannot be ignored. Therefore, the threshold voltage becomes dependent on the channel length for short-channel devices, which is called the *Short-Channel Effect* (SCE).[1]

---

[1]The term *short-channel effect* in literature is often used for various, different phenomena that occur in short-channel devices or sometimes it is used as a general term to comprise all behavior different from long-channel devices. However, in this text the term will be restricted to the sensitivity of the threshold voltage to the channel length (occasionally also called V_T roll-off).

In (2.9), the SCE term introduces a threshold voltage reduction through ΔV_T which has a strong dependence on the channel length. As L decreases, ΔV_T will increase, and in turn V_T will decrease [9].

Until now, it was possible to adjust the long-channel definition of the threshold voltage with additional terms to include short-channel behavior. However, if a channel is made very short, the source and drain depletion regions tend to approach each other in such a way that the long-channel behavior disappears and the short-channel behavior requires entirely different approaches [34]. To avoid this issue in modern advanced nanometer technologies, *halo* regions are introduced. Basically, halo implants are added near the ends of the channel, close to the source and drain. In these halo regions, the doping is locally increased to limit the depletion region widths around the source and the drain and to reduce SCE. In turn, this can cause a *Reverse* Short-Channel Effect (RSCE) where V_T will increase when channel length is reduced.

2.1.2.6 Remark

The concept of the threshold voltage inherently introduces some problems: reconciling gradual electric behavior of a transistor with one specific 'turn-on' point is bound to cause issues. Although a comprehensive theoretical definition (see Eq. (2.9)) was proposed in this section, in reality, it is difficult to define V_T unambiguously [27]. For instance, the value of parameter ϕ_0 is difficult to define exactly [34]. Another example is that in both the 90 nm and the 40 nm technologies at hand, V_T is simply defined as the gate-source voltage V_{gs} where I_{ds} equals a certain I_{ds,V_T}. This I_{ds,V_T} is only dependent on W/L multiplied by a constant technology-specific factor, which seems like a very rough simplification of the threshold voltage. Although V_T is hard to define in reality, this does not diminish the value of the definition discussed in this section as it provides the necessary insight in the different phenomena that influence the threshold voltage.

2.1.3 Region of Interest

This book focuses on ultra-low-voltage operation of digital circuits. Therefore, the region of interest of this work is the *sub-threshold* region where the diffusion current is dominant, as well as the *near-threshold* region where diffusion and drift currents are equally important. In the remainder of this book, the terms 'ultra-low-voltage' and 'sub-threshold' operation will be used to indicate the same concept, unless otherwise stated.

2.2 Challenges of Sub-Threshold Operation

Normal operation of a digital circuit means that such a circuit is operated at the nominal supply voltage $V_{dd,nom}$. In the technologies at hand, $V_{dd,nom}$ is 1 V for 90 nm CMOS and is 0.9 V for 40 nm CMOS. In the context of this book, this is considered nominal, *super-threshold* operation.

Sub-threshold circuits, on the other hand, operate at a supply voltage near or under the threshold voltage. Operating circuits in the sub-threshold region has the advantage of providing significant energy savings, as was discussed in Chap. 1. However, it also results in several issues which must be tackled to allow the widespread use of sub-threshold circuits. This section will discuss these different circuit-level challenges.

2.2.1 *Performance*

By operating a transistor at ultra-low supply voltages, the current I_{ds} of that transistor decreases significantly compared to nominal operation, as shown in Fig. 2.3. As a result, ultra-low-voltage circuits exhibit a large increase in delay because they use the weak inversion current as drive current. As an example, Fig. 2.7 displays the propagation delay t_p of a CMOS inverter which is sized for regular super-threshold operation as function of the supply voltage. Consequently, the operation frequency decreases due to the increase of the delay. In other words, sub-threshold circuits are only able to reach low to moderate circuit performance. Note that in the sub-threshold region, the delay is exponentially dependent on V_{dd}. In this region, a slightly higher supply voltage can thus result in a considerable speed increase, which is an important consideration to be made when designing sub-threshold circuits.

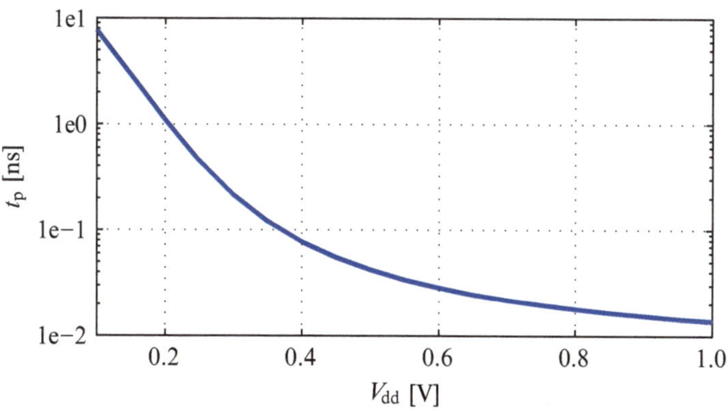

Fig. 2.7 Propagation delay of a regular-sized CMOS inverter ($W_{nMOS} = W_{min}$ and $W_{pMOS} = 3 \cdot W_{min}$) as function of V_{dd} (fan-out = 1)

Early papers on digital sub-threshold design were mainly focused on minimizing the energy consumption as much as possible. The speed of these circuits was only of secondary concern, thereby often resulting in kHz-performance, e.g. [7, 13, 15, 36]. Later on, sub-threshold research became more mature and more attention was paid on improving the circuit performance. Hence, some newer sub- or near-threshold designs report promising high ultra-low-voltage operating frequencies, well within the MHz-range, such as [5, 12, 14, 28]. However, there are still recent publications in advanced nanometer technologies which only reach kHz-performance, e.g. [17], or frequencies of a few MHz at higher supply voltages, such as [20, 25].

This book targets ultra-low-voltage circuits where MHz-performance is reached. Techniques to guarantee such speed performance are presented in Chaps. 3 and 4.

2.2.2 Leakage

Since ultra-low-voltage circuits use the exponential weak inversion current as their drive current, leakage plays a much bigger role in circuit functionality. More precisely, the difference between the on-current I_{on} (i.e. I_{ds} when $V_{gs} = V_{ds} = V_{dd}$) and the off-current I_{off} (i.e. I_{ds} when $V_{gs} = 0$ and $V_{ds} = V_{dd}$) reduces severely when V_{dd} is lowered. From a CMOS circuit perspective, the relevant I_{on}/I_{off} ratio is not the ratio between the currents of the same transistor, but rather the ratio between the currents of the nMOS transistor on one hand and the pMOS transistor on the other hand. A decrease in current ratio can cause unwanted leakage paths. The relevant current ratios thus provide a measure for the degradation of the functionality of a CMOS circuit when the supply voltage is reduced.

Figure 2.8 shows these current ratios as function of V_{dd}. In both technologies used throughout this book, the pMOS transistor is significantly weaker than the nMOS transistor in the weak inversion region. This can also be seen in Fig. 2.8 where the $I_{on,pMOS}/I_{off,nMOS}$ ratio is always approximately a factor 100 smaller than the $I_{on,nMOS}/I_{off,pMOS}$ ratio. At very low supply voltages, it becomes problematically low. Obviously, a ratio below 1 is dramatic because it implies that the drive current of the pMOS is lower than the leakage current of the nMOS. To avoid functional circuit failures, a value considerably higher than 1 is necessary. In this work, a value of 50 is considered the minimum feasible value for nominal current ratios, because constructing a functional logic gate with this ratio is still feasible.

To summarize, the *absolute* leakage current as such is not that important for the functionality of ultra-low-voltage circuits, it is the *relative* ratio between the on-current and the leakage or off-current which inserts a lower limit on the minimum possible supply voltage in order to guarantee circuit functionality. Although it is the relative ratio which is crucial for circuit robustness, the absolute leakage current of course still influences the circuit's power consumption.

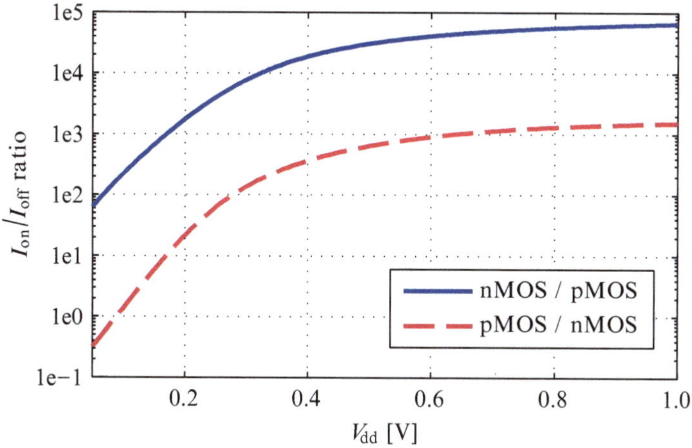

Fig. 2.8 Relevant nominal I_{on}/I_{off} ratios for minimal transistors as function of V_{dd}

2.2.3 *Variability*

Sub-threshold circuits suffer from an exponential sensitivity to variations due to the exponential current/voltage dependencies. This makes variability a critical, not to be neglected issue for sub-threshold operation. The increased sensitivity to variations compared to super-threshold circuits can severely compromise the robustness of circuits and the overall yield. Therefore, this section examines the different classifications of variations and their impact on ultra-low-voltage digital circuits.

The classical way of dealing with variations for digital circuits is to take design margins to ensure yield. However, because of the highly sensitive sub-threshold circuits, the accumulation of conservative design margins compromises the low-power benefit of operating in sub-threshold. Therefore, this is not the optimal manner of coping with the increased variations. This book focuses on *variation-resilient* circuit design as one of its main objectives to improve robustness without the need for an increased supply voltage or excessive design margins.

2.2.3.1 Inter-Die Variations

The first class is called *inter-die* variations, i.e. when variations on device parameters differ from one die to another. Technology foundries have quantified these inter-die variations by identifying *process corners*. The collective effect of process variations is lumped into their effect on nMOS and pMOS transistors [38]: *typical*, *fast* or *slow*. The different corners are then the combination of these effects, with the first abbreviation pointing to the nMOS and the second to the pMOS transistor. The typical-typical (or *tntp*) corner is the nominal case, while the *snsp* , *snfp* , *fnfp* and

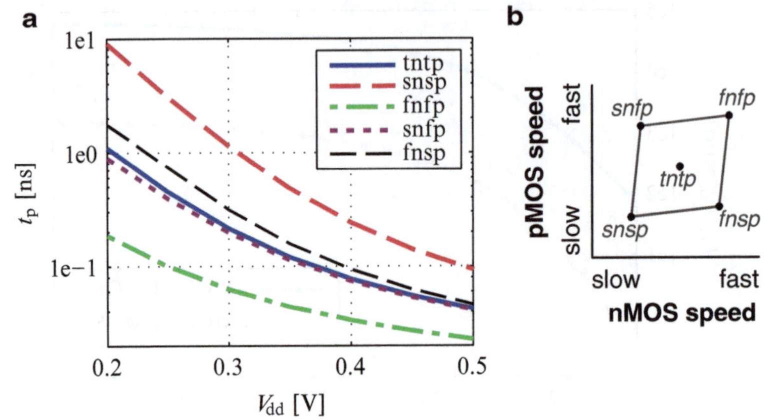

Fig. 2.9 (**a**) Propagation delay of a regular-sized CMOS inverter as function of V_{dd} in all process corners. (**b**) Visualization of the different process corners

fnsp characterize the extreme cases, as visualized in Fig. 2.9b. The imaginary box limited by the process corners is not square because some characteristics are shared by both nMOS and pMOS devices. To illustrate the impact of inter-die variations, the t_p of a regular-sized CMOS inverter is given as function of V_{dd} for the different process corners in Fig. 2.9a.

The snsp and fnfp corners mainly have an effect on the overall circuit speed. On the contrary, the snfp and the fnsp corners pose the highest threat on the reliability of ultra-low-voltage circuits since the relative speed and strength of the devices then exhibits the largest deviation.

2.2.3.2 Intra-Die Variations

Intra-die variations, on the other hand, classify variations on device parameters between transistors on the same die. *Monte Carlo* (MC) simulations are used to quantize the effect of intra-die variations. These variations are usually modeled with *normal* or *Gaussian* statistical distributions. For instance, Fig. 2.10 shows the Probability Density Function (PDF) of V_T of an nMOS transistor acquired with 1,000 MC simulations. The distribution is centered around the mean value μ, while the standard deviation σ provides a measure of the amount of dispersion or variation. Consequently, the standard deviation is directly tied to the yield of a circuit: if a circuit is functional taking into account 1σ variation, its yield is 68.27 %, while 3σ corresponds to a yield of 99.73 %. As visible in Fig. 2.10, σ_{V_T} is approximately 20 % of μ_{V_T}, which is quite a substantial factor especially considering the high susceptibility to variations of ultra-low-voltage circuits. Pelgrom has defined a coefficient which provides a measure of the amount of V_T variations of a certain technology, as will be discussed in Sect. 2.3.2.

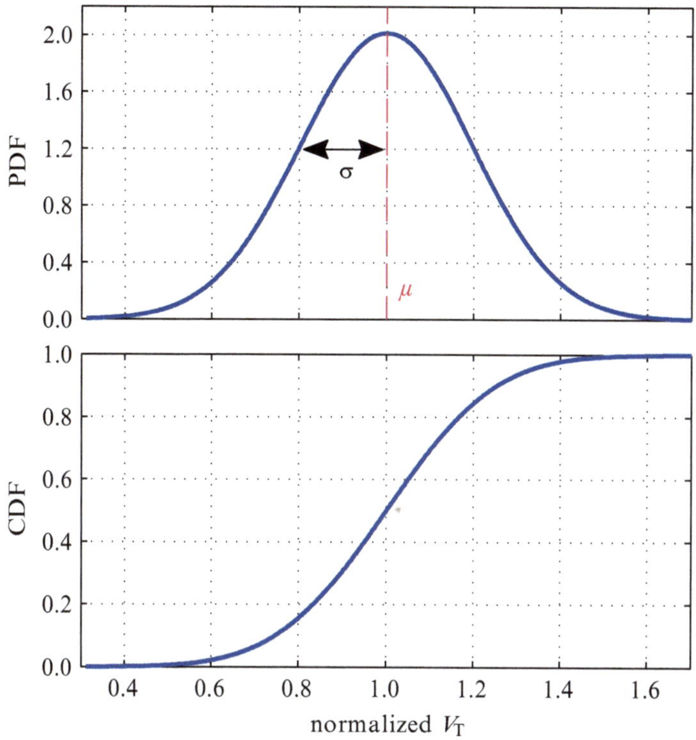

Fig. 2.10 PDF and CDF plots of the variation of V_T of an nMOS transistor, obtained with 1,000 MC simulations

To establish sub-threshold circuits as an attractive option for industrial applications, a high yield is essential. To design variation-resilient circuits, it is imperative to take into account the tails of the distribution as this is where the outliers are located which will determine the total yield. In this book, distributions will be visualized as Cumulative Distribution Functions (CDF) because they provide a better insight in the tails of the distribution than a PDF, as shown in Fig. 2.10.

Traditionally, intra-die variations were mostly an issue in analog design because of mismatch. Nowadays, however, it is also a key factor in digital design as variability increased considerably with CMOS scaling. On the one hand, intra-die variations impose an important threat on the reliability of ultra-low-voltage circuits since they result in a deteriorated functionality. On the other hand, gate delays become highly variable. To illustrate the latter, the percentage variation of t_p of a CMOS inverter under intra-die variations is shown in Fig. 2.11: when the supply reduces to low values, the variability aggravates significantly.

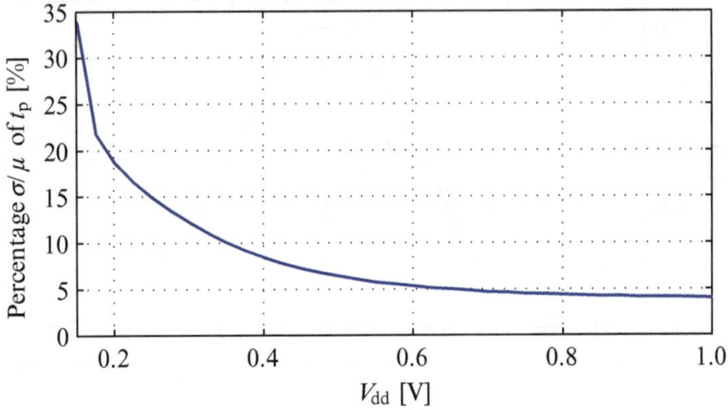

Fig. 2.11 Variation of the propagation delay of a regular-sized CMOS inverter as function of V_{dd}, obtained with 1,000 MC simulations

2.2.4 Temperature

Not only the value of the threshold voltage changes with temperature (see Sect. 2.1.2.3), the carrier mobility is influenced by temperature as well. The mobility μ and the threshold voltage V_T both decrease with temperature. The decreasing μ leads to a reduced current when the temperature is increased. The mobility effect is dominant at high supply voltages, resulting in slower circuits at higher temperatures, and vice versa. This temperature-dependence at nominal V_{dd} is well known. However, in the sub-threshold region, the lower V_T plays an important role as well since it causes I_{ds} to increase [37]. Both effects then counteract each other, and the resulting current increases slightly with temperature. Therefore, in the ultra-low-voltage domain, a higher temperature has a modest beneficial effect on the circuit performance [4].

Figure 2.12 shows the propagation delay of an inverter as function of temperature at a 200 mV supply voltage, for the industrial temperature range from −40 to 85 °C. As already reported in [4], negative Celsius temperatures are highly detrimental to ultra-low-voltage logic, which is clearly visible in Fig. 2.12. However, when going to a commercial temperature range between 0 and 70 °C, the impact of temperature is much less pronounced.

Moreover, ultra-low-voltage circuits are less subject to self-heating than high-performance circuits [18]. The question is then if the higher temperature limit of the commercial range of 70 °C is not excessive for ultra-low-voltage circuits. If they are operating in an environment with temperatures around room temperature, for instance between 15 and 40 °C, the temperature-dependence is rather limited in the sub-threshold region. Therefore, it has not been investigated more detailed for the circuits of this book. Nonetheless, it could be interesting to perform more

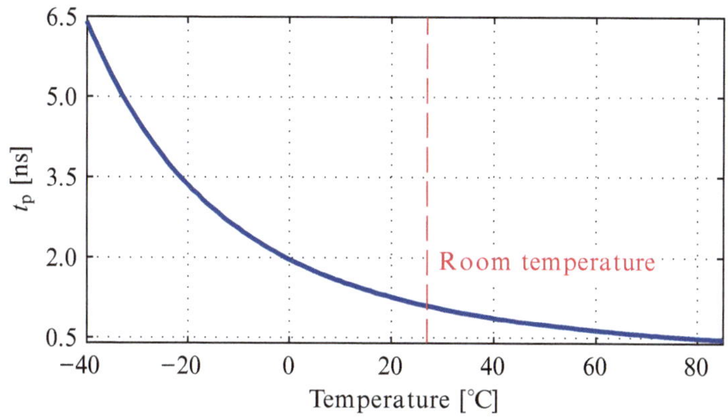

Fig. 2.12 Propagation delay of a regular-sized CMOS inverter as function of temperature (for $V_{dd} = 200\,\text{mV}$)

research on this temperature-dependence, as will be stated in Sect. 7.4. All presented simulation and measurement results of this book are performed at room temperature, unless otherwise stated.

2.3 Technology Scaling

The impact of CMOS technology scaling for digital circuits operating at the nominal supply has been extensively studied, both through simulations and through on-chip implementations and measurements. The influence of scaling on circuits operating in the weak inversion region has received some attention, but until now, this attention has been limited to device-level studies and circuit-level simulations [29].

Previous work investigated the effect of scaling on device-level and proposed different scaling strategies. In [10], a model study of sub-threshold transistors has been performed to investigate the implications of device scaling on sub-threshold operation from the 90 nm down to the 32 nm technology node. An alternative scaling strategy was proposed to help sub-threshold circuits to reliably scale to nanometer CMOS technologies. In [23], the focus was to redesign devices specifically for sub-threshold operation. An optimized transistor structure was proposed to improve the circuit delay and the Power-Delay Product (PDP) in the sub-threshold region. The impact of technology scaling for nodes from 90 nm to 22 nm was examined in [8] and strategies for increasing the robustness of sub-threshold circuits were proposed.

Some prior works also performed simulations to examine the impact of technology scaling on ultra-low-voltage logic circuits. It mostly consisted of simple circuits and in one case a more elaborate logic unit was simulated. In [33], simulations of a ring oscillator were used to validate an analytical approach for studying the effect of technology scaling and variability on performance of ultra-low-power integrated systems. The effects of process variations were exhaustively examined to study the sensitivity of a circuit in presence of these variations. In [2], the impact of technology scaling on sub-threshold logic was investigated in nodes from 0.25 μm to 32 nm CMOS. A circuit-level simulation of a benchmark 8-bit multiplier was used to study the scaling effects, first using Predictive Technology Models (PTM) and then validated by industrial models.

To the author's knowledge, only [11] presented measured results: two test chips were fabricated in a 130 nm and a 65 nm technology, consisting of a 1,000-stage inverter chain and a 41-stage ring oscillator, both operating at ultra-low voltages. The measurements of these two simple circuits were used to validate a body biasing technique to adaptively balance the pMOS and nMOS transistors in strength. No papers have been published that present the design and measurements of a full digital system, implemented in different CMOS technology nodes. Moreover, a significant amount of the previous work did not use industrial models, but rather relied on PTMs (e.g. [8, 10, 33]). PTMs are reasonably accurate transistor models that benchmark future generations of technologies and are therefore a useful resource for early circuit design research [39]. Although PTMs are very suitable to investigate scaling trends, simulating with these models does not provide the same value as designing with industrial models, followed by manufacturing and measuring the designed chip. This is especially true when operating in the weak inversion region with its exponential sensitivities.

One of the objectives of this book is to fill the hiatus between simulations and measured, confirmed results. Therefore, an extensive digital circuit has been designed, processed and measured in both a 90 nm and a 40 nm CMOS technology. The test vehicle that was used to study the effect of technology scaling on ultra-low-voltage circuits, is a 16-bit Multiply-Accumulate Unit (MAC). Chapter 5 will discuss the design and the measured results of the MAC in both technologies.

In the remainder of this section the fundamental limits of scaling as well as its expected effects will be discussed.

2.3.1 Fundamental Limits

Firstly, a practical expression that estimates the minimum feasible supply voltage that can be expected for digital circuits in a certain technology will be provided [29].

Previous research has focused on theoretically finding the fundamental limit for the lowest operating voltage for CMOS technologies. Already in 1972, [32] studied the minimum usable supply of an inverter, with the requirement that the inverter should have sufficient maximum gain at $V_{dd}/2$ to be usable in a digital circuit. Based on measurements in a technology available at that time, the authors estimated that the minimum usable V_{dd} would have a value of about 8 V_{th}, or 207 mV at 300 K.

In 2001, [6] proposed another theoretical limit of the lowest operable supply. To achieve this V_{dd}, the nMOS and pMOS off-currents must be equalized. Following this requirement, the ideal supply limit of 4 V_{th} is proposed, which is 103 mV at 300 K.

These are all theoretical limits that predict the lowest possible supply voltage of CMOS digital circuits. However, they are not practical limits that take into account the specific details of the technology at hand. Therefore, a practical limit for the minimum feasible supply will be derived from the equations listed below. The basic Eq. (2.8) for the current flow in the weak inversion region consists of an exponential relationship:

$$I_{ds} = I_0 \cdot \exp\left(\frac{V_{gs} - V_T}{n \cdot V_{th}}\right)$$

The on-current I_{on} at $V_{gs} = V_{dd}$ and the off-current I_{off} at $V_{gs} = 0$ can thus be derived from (2.8):

$$I_{on} = I_0 \cdot \exp\left(\frac{V_{dd} - V_T}{n \cdot V_{th}}\right) \quad (2.14)$$

$$I_{off} = I_0 \cdot \exp\left(\frac{-V_T}{n \cdot V_{th}}\right) \quad (2.15)$$

Taking the ratio of (2.14) and (2.15) gives:

$$\frac{I_{on}}{I_{off}} = \frac{I_0 \cdot \exp\left(\frac{V_{dd}}{n \cdot V_{th}}\right) \cdot \exp\left(\frac{-V_T}{n \cdot V_{th}}\right)}{I_0 \cdot \exp\left(\frac{-V_T}{n \cdot V_{th}}\right)}$$

$$= \exp\left(\frac{V_{dd}}{n \cdot V_{th}}\right) \quad (2.16)$$

In (2.16), a direct relationship between the variables V_{dd}, I_{on} and I_{off} is obtained since V_{th} is fixed for a certain temperature and n is fixed for a certain technology. An equation for the supply voltage can be derived:

$$V_{dd} = \ln\left(\frac{I_{on}}{I_{off}}\right) \cdot n \cdot V_{th} \quad (2.17)$$

The value of n is affected by the depletion region characteristics [34]:

$$n = 1 + \frac{C_D}{C_{ox}} \quad (2.18)$$

where C_D is the depletion layer capacitance and C_{ox} is the gate oxide capacitance. The value of n is equal to 1 for an ideal transistor, but unfortunately larger than 1 for actual devices. It is typically in the range of 1.3–1.7 for CMOS processes [38]. Since it is difficult to accurately determine n for a certain technology, the link with the so-called sub-threshold slope S_S will be made. The sub-threshold slope is defined by the amount by which V_{gs} must be increased in order for the weak inversion current I_{ds} to be increased by one order of magnitude, and it is expressed in mV/decade. The sub-threshold slope is expressed as: [34]

$$S_S = n \cdot V_{th} \cdot \ln(10) \tag{2.19}$$

Substituting n in (2.17) by using (2.19), results in:

$$\begin{aligned} V_{dd} &= \ln\left(\frac{I_{on}}{I_{off}}\right) \cdot \frac{S_S}{V_{th} \cdot \ln(10)} \cdot V_{th} \\ &= \log_{10}\left(\frac{I_{on}}{I_{off}}\right) \cdot S_S \end{aligned} \tag{2.20}$$

This result shows that for a certain CMOS technology (and thus for a certain S_S), the minimum supply V_{dd} is only dependent on the minimum I_{on}/I_{off} current ratio. This equation makes it possible to derive a practical as well as a theoretical limit for the minimum feasible supply voltage for a circuit operating in the weak inversion region. The supply-dependence of the I_{on}/I_{off} ratio is logical: the lower V_{dd}, the lower I_{on} will be obtained, and the lower the current ratio will become. From experience, a fair minimum value for the I_{on}/I_{off} current ratio is 50. A lower value of the current ratio becomes problematic, since the circuit robustness in the presence of variations will be compromised. A theoretical limit for the minimum supply voltage can be found through the theoretical lower bound of the sub-threshold slope S_S. In the ideal case, n is equal to 1 and therefore the minimum S_S is equal to 60 mV/decade at room temperature. The theoretical $V_{dd,min}$ can then be calculated with (2.20) to be 101 mV.

However, although devices that have an ideal sub-threshold slope are optimal for sub-threshold applications [16], typical S_S values for a bulk CMOS process range from 70 to 120 mV/decade [30], well above the theoretical lower bound. Unfortunately, CMOS technology scaling has a bad impact on the sub-threshold slope because S_S is proportional to the gate oxide thickness t_{ox} which does not scale in proportion to the physical gate length. Scaling of t_{ox} actually slows down starting from the 130 nm node to limit gate leakage [2]. A comparison of industrial publications in [39] indicated that t_{ox} has been reduced by a mere 10 % per generation between the 130 nm and the 40 nm technology nodes [10]. For the technologies at hand, t_{ox} decreases even less with about 13 % between the 90 nm and 40 nm nodes. As a result, S_S degrades as function of CMOS technology scaling.

In the near future, different process technologies can alter this deteriorating trend. For instance, transistors in Silicon-On-Insulator (SOI) technologies have the advantage of a near-ideal sub-threshold slope, e.g. fully depleted SOI transistors have an S_S of 65–85 mV/decade [35]. Another improvement would be to use a bulk

Table 2.1 Measured sub-threshold slope S_S and resulting $V_{dd,min}$ [29]. Bold values indicate the practical limits per technology

CMOS Technology		90 nm	40 nm
S_S	[mV/decade]		
– nMOS		93.0	98.1
– pMOS		86.1	109.9
$V_{dd,min}$	[mV]		
– nMOS		**158**	166
– pMOS		146	**186**

CMOS technology with a high-K dielectric instead of the conventional SiO_2 gate oxide ($K_{SiO_2} = 3.9$, see Eq. (2.3)). As stated before, scaling of t_{ox} is slowing down. This is because of the rapid increase in gate leakage current due to tunneling. Using a dielectric with a higher K allows to increase the gate oxide capacitance for the same oxide thickness, or to keep C_{ox} unchanged and use a larger t_{ox}. The latter option would reduce the tunneling effects [21] and therefore the induced leakage and would also increase S_S and therefore the I_{on}/I_{off} current ratio for a given V_{dd}.

However, this work concentrates on ultra-low-voltage design in bulk CMOS technologies with a conventional SiO_2 gate oxide. Measurements of transistors in both CMOS technologies at hand confirm the deteriorating trend (see Table 2.1): S_S degrades 5.5 % going from the 90 nm to the 40 nm technology for an nMOS transistor and 27.6 % for a pMOS. The practical limit of $V_{dd,min}$ can then be calculated as the maximum $V_{dd,min}$ per technology, resulting in 158 mV for the 90 nm technology and 186 mV for the 40 nm technology, as indicated in bold.

To conclude, by using a practical limit for the I_{on}/I_{off} current ratio and a value of S_S (which can be obtained through simulations or measurements), a straightforward manner is obtained to calculate the practical minimum feasible supply voltage for digital circuits operating in the weak inversion region in a certain CMOS technology.

In Sect. 5.2.4, this equation to calculate the practical limit of $V_{dd,min}$ will be validated with measurement results of an identical system fabricated in two CMOS technology nodes. As will be seen, this practical limit predicts the measured values quite well.

The goal of this equation is to provide a fast but reasonably accurate estimation for $V_{dd,min}$ without having to resort to large amounts of simulations or measurements. However, in literature, some papers have performed interesting work on the latter. For instance, [19] proposes a method to extract minimum supply voltage and failure rate of digital gates using static noise margins. In [1], it was shown that the increase of DIBL in advanced nanometer technologies results in an increase of the minimum feasible supply voltage. In [22] extensive measurements of different lengths of ring oscillators verify that the lower limit of V_{dd} increases considerably with increasing number of stages.

2.3.2 Impact of Scaling

This section investigates what the expected impact is that technology scaling has on ultra-low-voltage designs [29]. In this book, two CMOS technologies are used for the design of the prototypes, i.e. 90 nm and 40 nm standard CMOS. The following analysis will therefore focus on these two technologies.

With technology scaling from 90 nm to 40 nm CMOS, both the nominal supply voltage V_{dd} and the transistor threshold voltage V_T decrease slightly. Since V_T decreases and the W_{min}/L_{min} ratio does not remain constant but instead increases with a factor of 2 (for the technologies at hand), one would expect the transistor current I_{ds} to increase for a certain supply. Therefore, the delay should reduce, causing circuits to function at a higher speed. The dynamic energy consumption is expected to decrease with the third power of the scaling factor. The total chip area will decrease, as well as the cost. These are all advantages of scaling. However, there are also downsides to scaling. First, the leakage current will increase exponentially due to the reduced V_T. Second, because of the reduced transistor dimensions, transistor variability will have a higher impact and will thus become much more important.

When looking at technology scaling from an ultra-low-voltage perspective, all of the advantages remain. In fact, because of the decreasing V_T, circuits should become much faster for the same extremely low supply voltage. Since speed performance is often an issue in sub- and near-threshold designs, technology scaling becomes an even more attractive option for such designs. However, the disadvantages of scaling have an even higher effect in ultra-low-voltage designs. Due to the increased leakage and the aforementioned degradation of sub-threshold slope (see Sect. 2.3.1), the I_{on}/I_{off} ratios in the sub-threshold domain reduce to dramatically low values, thereby compromising circuit robustness. Moreover, the exponential sensitivity to variations combined with the overall increased variability results in problematic gate robustness. To conclude, technology scaling can definitely bring added value for ultra-low-voltage circuits, but it is imperative to take into account the low current ratios and the high variability at the time of design.

To obtain an understanding of the increased variations, a look at the impact of technology on threshold voltage variation will be given. The threshold voltage is determined by the number and location of dopant atoms implanted in the channel region. Since the number of dopants is small in nanometer processes, the variation of V_T due to random dopant fluctuations becomes large [38]. The Pelgrom coefficient A_{V_T} [24] provides a measure of the amount of V_T variations:

$$\sigma_{V_T} = \frac{A_{V_T}}{\sqrt{W \cdot L}} \tag{2.21}$$

The Pelgrom coefficient for both technologies has been extracted out of 1,000 MC transistor simulations. For nMOS transistors, A_{V_T} increases with 5.6 % when going from 90 nm to 40 nm, while the increase for pMOS transistors is as large as 41.4 %. These numbers clearly show the increased variations for advanced nanometer technologies.

2.3.3 Model Accuracy

An additional problem in the weak inversion region, is that the calibration of the transistor models is not as reliable as it is in the strong inversion, nominal region [29]. To give an example, Fig. 2.13a shows the simulation results of the on-current I_{on} of a minimal nMOS as function of V_{dd}. According to the foundry provided models of the 90 nm and 40 nm CMOS technologies at hand, $I_{on,40\,nm}$ is more than two times smaller than $I_{on,90\,nm}$ for a supply of 200 mV, which directly contradicts the scaling effects that were expected. A similar unexpected behavior can be seen in Fig. 2.13b, which shows the propagation delay t_p of a regular-sized inverter as function of V_{dd}. For the same 200 mV supply, $t_{p,40\,nm}$ is 56 % higher than $t_{p,90\,nm}$, resulting in an inverter which is more than 1.5 times slower. Figure 2.13 also shows that in the nominal supply domain on the contrary, the technology scaling expectations do apply. To conclude, these simulation results show that blindly relying on transistor models in the ultra-low-voltage domain is not recommendable.

That model accuracy differs between the strong and the weak inversion region is also indicated by the technology foundries. The accuracy criteria vary greatly between these regions: e.g. for 90 nm CMOS, the requirement to ensure model quality is that the fitting of the simulation model compared to the measured characteristics should have an error less than 75 % for I_{ds} of a transistor operating in the weak inversion region, while this requirement decreases to only 7 % when the transistor is operating in strong inversion.

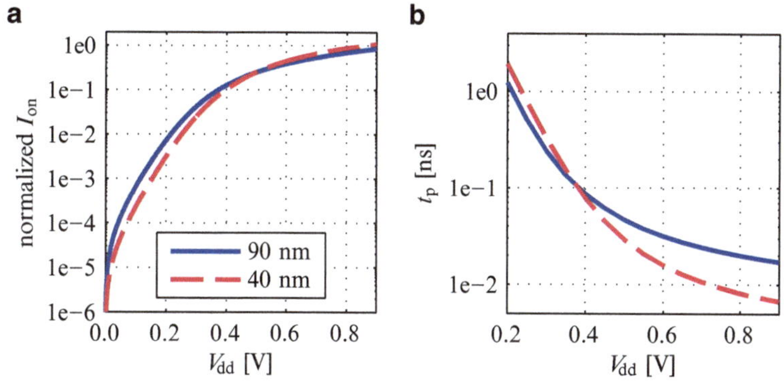

Fig. 2.13 Technology model comparison: (**a**) on-current for a minimally sized nMOS (normalized to $I_{on,90\,nm}$ at $V_{dd,nom}$) and (**b**) propagation delay for a regular-sized inverter (fan-out = 1)

Some nuances should be given to the discussion of the reliability of the transistor models. Although *absolute* numbers are not very trustworthy (such as the exact energy consumption or propagation delay of a circuit), *relative* comparisons in the same technology using the simulation results are definitely valuable. Out of experience from the designs described in this work (see Chaps. 5 and 6), we also know that simulations to check functionality produce reliable results. Moreover, as can be seen from the variation analysis above, the intra-die simulations do show the expected increase in variability when going to a smaller technology and therefore such simulations do provide realistic results.

2.4 Transistor Type

Around the 130 nm technology node, foundries began to offer on the one hand high-performance (sometimes also called general purpose or standard performance) and on the other hand low-leakage technology options. These process options were introduced in order to be able to optimize circuits according to whether they are more speed-constrained or more energy-constrained. For some technologies, it is possible to use the two options on the same wafer, but in most technologies it is only possible to choose one single option for a specific wafer.

Furthermore, modern deep sub-micron or advanced nanometer CMOS technologies also offer multiple V_T-options, and therefore a choice has to be made concerning the threshold voltage selection. Typically, a certain technology offers three different transistor types:

- Low-V_T or LVT transistors have the lowest threshold voltage of the three types. As a result, these transistors exhibit the highest speed, at the penalty of a higher leakage current. They are mostly used in cases where timing is critical and a high speed is needed.
- High-V_T or HVT transistors have the highest threshold voltage of the three types. These transistors have the lowest leakage current of the three types, at the penalty of a lower speed. They are used in cases where timing is not critical and leakage power can thus be reduced.
- Standard-V_T or SVT transistors have an intermediate threshold voltage and thus medium speed and leakage characteristics compared to the other types.

The inherent disadvantage of working in the sub-threshold region is the speed deterioration. However, by using LVT transistors that have higher currents than SVT or HVT transistors for the same supply voltage, a maximal sub-threshold speed can be guaranteed. Naturally, the use of LVT transistors also aggravates leakage. However, the increased leakage can be handled by taking this into account during circuit-level design. This will be explained in detail in Chap. 3. The building blocks in this book are therefore always constructed with LVT transistors, which is a first step to obtain sub-threshold circuits with MHz-range operating frequencies. The same reasoning also holds for why the chips in this book are always designed in the high-performance flavor of a CMOS technology, and not in the low-leakage flavor.

Similarly, in [3], it has been shown that using the high-performance option of a technology is beneficial for operation at minimal energy consumption for frequencies in the MHz-range, while the low-leakage option should be favored for kHz-range operation.

Simulation results in this text are therefore always obtained with LVT transistors in the high-performance process of a technology, unless stated otherwise.

2.5 Conclusion

This chapter discussed various important aspects to fully grasp sub-threshold operation before exploring ultra-low-voltage circuit design in the following chapters. The exponential behavior of transistors operating in the weak inversion region is examined, as well as its dependence on the threshold voltage. The inherently low to moderate performance of ultra-low-voltage circuits, the reduced current ratios, the high sensitivity to both inter- and intra-die variations, as well as the influence of temperature are the main challenges of sub- or near-threshold circuit design and were therefore examined in detail in this chapter.

In this book, prototypes in two different technologies were fabricated, i.e. in 90 nm and 40 nm CMOS. In the interest of understanding the differences between both technologies, this chapter analyzed the impact of technology scaling, together with the fundamental limits scaling sets to sub-threshold circuits. To conclude, a short examination of the different transistor types offered by modern CMOS technologies, resulted in the recommended use of LVT transistors because of their ability to offer higher currents for the same supply compared to the other transistor types.

References

1. Bol D, Ambroise R, Flandre D, Legat JD (2008) Analysis and minimization of practical energy in 45nm subthreshold logic circuits. In: Proceedings of the IEEE international conference on computer design (ICCD), pp 294–300. DOI: 10.1109/ICCD.2008.4751876
2. Bol D, Ambroise R, Flandre D, Legat JD (2009) Interests and limitations of technology scaling for subthreshold logic. IEEE Trans Very Large Scale Integ (VLSI) Syst 17(10):1508–1519. DOI: 10.1109/TVLSI.2008.2005413
3. Bol D, Flandre D, Legat JD (2009) Technology flavor selection and adaptive techniques for timing-constrained 45 nm subthreshold circuits. In: Proceedings of the ACM/IEEE international symposium on low power electronics and design (ISLPED), pp 21–26
4. Bol D, Hocquet C, Flandre D, Legat JD (2010) The detrimental impact of negative celsius temperature on ultra-low-voltage CMOS logic. In: Proceedings of the IEEE European solid-state circuits conference (ESSCIRC), pp 522–525. DOI: 10.1109/ESSCIRC.2010.5619758

5. Bol D, De Vos J, Hocquet C, Botman F, Durvaux F, Boyd S, Flandre D, Legat JD (2013) Sleepwalker: A 25-MHz 0.4-V sub-mm^2 7-μW/MHz microcontroller in 65-nm LP/GP CMOS for low-carbon wireless sensor nodes. IEEE J Solid State Circuits 48(1):20–32. DOI: 10.1109/JSSC.2012.2218067
6. Bryant A, Brown J, Cottrell P, Ketchen M, Ellis-Monaghan J, Nowak E (2001) Low-power CMOS at Vdd = 4kT/q. In: Proceedings of the IEEE device research conference (DRC), pp 22–23. DOI: 10.1109/DRC.2001.937856
7. Calhoun B, Chandrakasan A (2006) Ultra-dynamic voltage scaling (UDVS) using sub-threshold operation and local voltage dithering. IEEE J Solid State Circuits 41(1):238–245. DOI: 10.1109/JSSC.2005.859886
8. Calhoun B, Khanna S, Mann R, Wang J (2009) Sub-threshold circuit design with shrinking CMOS devices. In: Proceedings of the IEEE international symposium on circuits and systems (ISCAS), pp 2541–2544. DOI: 10.1109/ISCAS.2009.5118319
9. Cheng Y, Chan M, Hui K, Jeng MC, Liu Z, Huang J, Chen K, Chen J, Tu R, Ko PK, Hu C (1996) BSIM3v3 manual. Department of Electrical Engineering and Computer Sciences, University of California, Berkeley
10. Hanson S, Seok M, Sylvester D, Blaauw D (2008) Nanometer device scaling in subthreshold logic and SRAM. IEEE Trans Electron Devices 55(1):175–185. DOI: 10.1109/TED.2007.911033
11. Hwang ME (2011) Supply-voltage scaling close to the fundamental limit under process variations in nanometer technologies. IEEE Trans Electron Devices 58(8):2808–2813. DOI: 10.1109/TED.2011.2151257
12. Jeon D, Seok M, Chakrabarti C, Blaauw D, Sylvester D (2012) A super-pipelined energy efficient subthreshold 240 MS/s FFT core in 65 nm CMOS. IEEE J Solid State Circuits 47(1):23–34. DOI: 10.1109/JSSC.2011.2169311
13. Kao J, Miyazaki M, Chandrakasan A (2002) A 175-mV multiply-accumulate unit using an adaptive supply voltage and body bias architecture. IEEE J Solid State Circuits 37(11):1545–1554. DOI: 10.1109/JSSC.2002.803957
14. Kaul H, Anders M, Mathew S, Hsu S, Agarwal A, Krishnamurthy R, Borkar S (2008) A 320mV 56μW 411GOPS/Watt ultra-low voltage motion estimation accelerator in 65 nm CMOS. In: Proceedings of the IEEE international solid-state circuits conference (ISSCC), pp 316–317. DOI: 10.1109/ISSCC.2008.4523184
15. Kim C, Soeleman H, Roy K (2003) Ultra-low-power DLMS adaptive filter for hearing aid applications. IEEE Trans Very Large Scale Integr VLSI Syst 11(6):1058–1067. DOI: 10.1109/TVLSI.2003.819573
16. Kim JJ, Roy K (2004) Double gate-MOSFET subthreshold circuit for ultralow power applications. IEEE Trans Electron Devices 51(9):1468–1474. DOI: 10.1109/TED.2004.833965
17. Konijnenburg M, Cho Y, Ashouei M, Gemmeke T, Kim C, Hulzink J, Stuyt J, Jung M, Huisken J, Ryu S, Kim J, de Groot H (2013) Reliable and energy-efficient 1MHz 0.4V dynamically reconfigurable SoC for ExG applications in 40nm LP CMOS. In: Proceedings of the IEEE international solid-state circuits conference (ISSCC), pp 430–431. DOI: 10.1109/ISSCC.2013.6487801
18. Kumar R, Kursun V (2007) Temperature-adaptive energy reduction for ultra-low power-supply-voltage subthreshold logic circuits. In: Proceedings of the IEEE international conference on electronics, circuits and systems (ICECS), pp 1280–1283. DOI: 10.1109/ICECS.2007.4511231
19. Kwong J, Chandrakasan A (2006) Variation-driven device sizing for minimum energy subthreshold circuits. In: Proceedings of the ACM/IEEE international symposium on low power electronics and design (ISLPED), pp 8–13. DOI: 10.1109/LPE.2006.4271799
20. Lutkemeier S, Jungeblut T, Berge H, Aunet S, Porrmann M, Ruckert U (2013) A 65nm 32b subthreshold processor with 9T multi-Vt SRAM and adaptive supply voltage control. IEEE J Solid State Circuits 48(1):8–19. DOI: 10.1109/JSSC.2012.2220671
21. Narendra S, Chandrakasan A (2006) Leakage in nanometer CMOS technologies. Springer, Heidelberg

22. Niiyama T, Piao Z, Ishida K, Murakata M, Takamiya M, Sakurai T (2008) Increasing minimum operating voltage (VDDmin) with number of CMOS logic gates and experimental verification with up to 1Mega-stage ring oscillators. In: Proceedings of the ACM/IEEE international symposium on low power electronics and design (ISLPED), pp 117–122. DOI: 10.1145/1393921.1393952
23. Paul B, Raychowdhury A, Roy K (2005) Device optimization for digital subthreshold logic operation. IEEE Trans Electron Devices 52(2):237–247. DOI: 10.1109/TED.2004.842538
24. Pelgrom M, Duinmaijer ACJ, Welbers A (1989) Matching properties of MOS transistors. IEEE J Solid State Circuits 24(5):1433–1439. DOI: 10.1109/JSSC.1989.572629
25. Pu Y, Pineda de Gyvez J, Corporaal H, Ha Y (2010) An ultra-low-energy multi-standard JPEG co-processor in 65 nm CMOS with sub/near threshold supply voltage. IEEE J Solid State Circuits 45(3):668–680. DOI: 10.1109/JSSC.2009.2039684
26. Rabaey J, Chandrakasan A, Nikolic B (2003) Digital integrated circuits: a design perspective, 2nd edn. Prentice Hall, Upper Saddle River
27. Razavi B (2001) Design of analog CMOS integrated circuits, 2nd edn. McGraw-Hill, Boston
28. Reynders N, Dehaene W (2014) A 210mV 5MHz variation-resilient near-threshold JPEG encoder in 40nm CMOS. In: Proceedings of the IEEE international solid-state circuits conference (ISSCC), pp 456–457
29. Reynders N, Dehaene W (2015) On the effect of technology scaling on variation-resilient subthreshold circuits. Elsevier Solid State Electron 103:19–29
30. Roy K, Mukhopadhyay S, Mahmoodi-Meimand H (2003) Leakage current mechanisms and leakage reduction techniques in deep-submicrometer CMOS circuits. Proc IEEE 91(2):305–327. DOI: 10.1109/JPROC.2002.808156
31. Sansen W (2006) Analog design essentials. Springer, New York
32. Swanson R, Meindl J (1972) Ion-implanted complementary MOS transistors in low-voltage circuits. IEEE J Solid State Circuits 7(2):146–153. DOI: 10.1109/JSSC.1972.1050260
33. Tajalli A, Leblebici Y (2011) Design trade-offs in ultra-low-power digital nanoscale CMOS. IEEE Trans Circuits Syst Regul Pap 58(9):2189–2200. DOI: 10.1109/TCSI.2011.2112595
34. Tsividis Y, McAndrew C (2011) Operation and modeling of the MOS transistor, 3rd edn. Oxford University Press, Oxford
35. Vitale S, Wyatt P, Checka N, Kedzierski J, Keast C (2010) FDSOI process technology for subthreshold-operation ultralow-power electronics. Proc IEEE 98(2):333–342. DOI: 10.1109/JPROC.2009.2034476
36. Wang A, Chandrakasan A (2005) A 180-mV subthreshold FFT processor using a minimum energy design methodology. IEEE J Solid State Circuits 40(1):310–319. DOI: 10.1109/JSSC.2004.837945
37. Wang A, Calhoun B, Chandrakasan A (2006) Sub-threshold design for ultra low-power systems. Springer, New York
38. Weste N, Harris D (2011) CMOS VLSI design: a circuits and systems perspective, 4th edn. Addison-Wesley, New York
39. Zhao W, Cao Y (2006) New generation of predictive technology model for sub-45nm design exploration. In: Proceedings of the IEEE international symposium on quality electronic design (ISQED), pp 585–590. DOI: 10.1109/ISQED.2006.91

Chapter 3
Gate-Level Building Blocks

This chapter discusses the gate-level building blocks which have been used to design the ultra-low-voltage prototypes of this work. Their aim was not only to operate at very low supply voltages in a variation-resilient manner, but also to function at speeds of $n \times 10\,\text{MHz}$. Such targets are only possible to achieve when attention is paid to both the transistor-level basic circuits and the architectural level (to be discussed in Chap. 4).

Careful design of logic gates is crucial if they should be able to efficiently work in the ultra-low-voltage region. Their topology not only has a large impact on the variation-resilience of the total design, but also on the delay, leakage power and active energy consumption. Therefore, Sect. 3.1 provides an elaborate comparison of circuit topologies, from very common logic families to more exotic circuit topologies which have been specifically proposed for operation in the ultra-low-voltage region [33]. An in-depth analysis of the characteristics of these logic families leads to the presentation of the circuit topologies that are preferred in this work in Sect. 3.2.

Section 3.3 continues this discussion of basic building blocks by exploring various memory elements. Not only their functionality differences are examined, but the trade-offs that accompany operation at low supply voltages as well.

To conclude, a summary of the different sizing options of the basic building blocks which have been employed in the four prototypes that will be presented in Chaps. 5 and 6 is given by Sect. 3.4. Finally, Sect. 3.5 ends this chapter.

N. Reynders, W. Dehaene, *Ultra-Low-Voltage Design of Energy-Efficient Digital Circuits*, Analog Circuits and Signal Processing,
DOI 10.1007/978-3-319-16136-5_3

3.1 Circuit Topology Comparison

To implement a certain logic function, there exist numerous possible circuit topologies. Several characteristics are important to evaluate the quality of a logic gate: speed, dynamic energy, leakage power, variation-resilience, robustness and area. This section discusses various topologies and determines their suitability for use with ultra-low supply voltages [33].

The standard value of the supply voltage at which circuit topologies will be evaluated in this comparison is 200 mV, unless stated otherwise. The analysis will be performed for the 90 nm CMOS technology at hand. However, the sizings and trade-offs are very similar for the 40 nm CMOS technology used in this work. In case of large differences, they will be explained.

3.1.1 Standard CMOS Logic

3.1.1.1 Concept

Figure 3.1 shows a generic implementation of an n-input *standard CMOS* logic gate. A standard CMOS gate is a combination of two complimentary networks: a Pull-Down Network (PDN) and a Pull-Up Network (PUN). The PDN provides a connection between the output and the ground when the logic function of the inputs is such that the output should be a logic '0'. It consists solely of nMOS transistors because incorporating pMOS transistors would result in V_T loss. The PUN, on the other hand, provides a connection between the output and the supply rail when the logic function of the inputs is such that the output should be a logic '1'. Equivalently, the PUN only consists of pMOS transistors. The networks are arranged so that for any input pattern, one of them will be on and the other one will be off. The inputs are connected to the gates of the nMOS and pMOS transistors. Important for the circuit topology comparison is that standard CMOS logic gates are inherently inverting. The PDN turns on when the inputs are '1', leading to '0' at the output, and vice versa. It is therefore not possible to realize a non-inverting Boolean function with standard CMOS logic gates in a single stage.

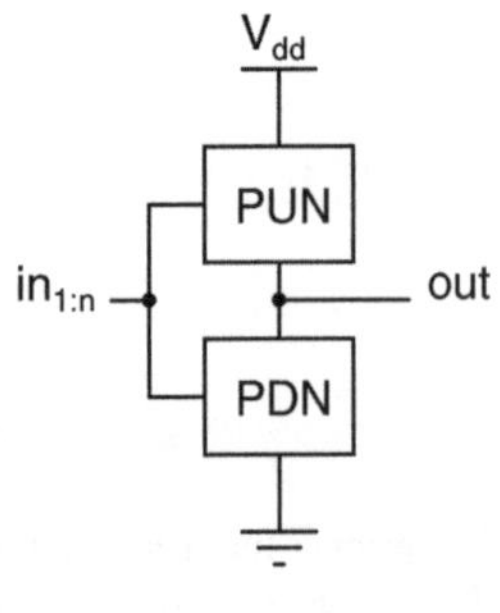

Fig. 3.1 Generic implementation of an n-input standard CMOS logic gate

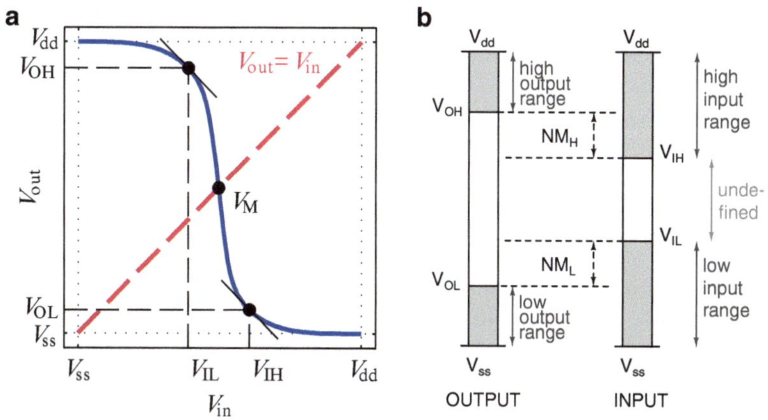

Fig. 3.2 Important DC characteristics of an inverter: (**a**) voltage transfer characteristic and (**b**) definition of noise margins

3.1.1.2 Ultra-Low-Voltage Operation

Several characteristics of logic gates will be used to adequately compare the operation of these gates at low supply voltages. The most basic logic gate is an inverter, which is implemented with a single pMOS transistor in the PUN and a single nMOS in the PDN of Fig. 3.1. The characteristics of an inverter provide an excellent measure of the quality of a certain circuit topology for ultra-low-voltage operation. Therefore, this text first discusses several properties of an inverter. If a comparison based on these properties does not suffice, other more complicated logic gates will be taken into account for the analysis.

First, the DC characteristics will be discussed in detail. Figure 3.2a shows the Voltage Transfer Characteristic (VTC) of an inverter. The VTC plots the output voltage V_{out} as function of the input voltage V_{in}. A first property which can be derived from the VTC is the *switching threshold voltage* of an inverter V_M. It can be found graphically at the intersection of the VTC curve and the line with function $V_{out} = V_{in}$. V_M provides a measure of the gate's symmetry: if V_M is equal to $V_{dd}/2$, the gate is *unskewed*. Otherwise, the gate is low or high *skewed*. In general, an unskewed gate is desired, as this provides maximal noise margins. However, a gate is sometimes intentionally skewed if more noise is expected on one of the logic levels, or to save area.

The *noise margin* is a measure of the sensitivity of a gate to noise [31]: the low noise margin NM_L and the high noise margin NM_H provide the maximal allowable noise level that a logic gate can withstand so that the input will still be interpreted correctly. The noise margins can be calculated through different points on the VTC in Fig. 3.2a, where the gain of the inverter equals -1. These unity gain points provide the minimum high and maximum low input and output voltages V_{IH} and V_{IL}, and V_{OH} and V_{OL}, respectively. Figure 3.2b visualizes the calculation of the noise margins for cascaded gates. NM_{L} is defined as the difference between V_{IL} and V_{OL}, while NM_{H} is the difference between V_{OH} and V_{IH}:

$$NM_{\mathrm{L}} = V_{\mathrm{IL}} - V_{\mathrm{OL}} \tag{3.1}$$

$$NM_{\mathrm{H}} = V_{\mathrm{OH}} - V_{\mathrm{IH}} \tag{3.2}$$

The region between V_{IH} and V_{IL} is called the undefined region because it does not represent a valid digital logic level. Evidently, the noise margins should be larger than 0 to obtain a functional digital circuit. The higher the noise margins, the lower the gate's sensitivity to noise. An unskewed gate has equal noise margins, which maximizes immunity to arbitrary noise sources [46]. In the ultra-low-voltage perspective, equal noise margins allow for operation at the lowest supply voltage, making it very desirable to have balanced noise margins.

The *gain* of the inverter plays an important role as well. The gain is the slope $\mathrm{d}V_{\mathrm{out}}/\mathrm{d}V_{\mathrm{in}}$ of the VTC. The gain defines whether the logic gate is *regenerative*. Regeneration signifies that a signal that deviates from the nominal levels V_{OL} or V_{OH} gradually converges back to those levels after passing through a number of such logic gates. In order for a gate to be regenerative, it has to satisfy some conditions: in the undefined region, the absolute value of the gain should be higher than 1, while it should be less than 1 in the valid regions. Note that this last requirement regarding the low gain regions thus directly implies positive noise margins.

Second, the transient characteristics will be discussed. Figure 3.3 gives the definition of the various delays of an inverter, and by extension for any logic gate. The *rise* time t_{r} and the *fall* time t_{f} provide a metric of the slopes of a waveform. They express with which delay a signal transits between different signal levels, and are defined by their transitions through 10 and 90 % of V_{dd}.

The *propagation delay* t_{p} is the time required for a signal to travel from the input of a logic gate to its output. It is measured between the 50 % transition points of the input and output waveforms (Fig. 3.3). Because a gate responds differently depending on whether it concerns a rising or a falling input transition, t_{pLH} and t_{pHL} differentiate between both such delays. The propagation delay t_{p} is then defined as the average of t_{pLH} and t_{pHL}:

$$t_{\mathrm{p}} = \frac{t_{\mathrm{pLH}} + t_{\mathrm{pHL}}}{2} \tag{3.3}$$

Naturally, these DC and transient characteristics will be compared not only for nominal operation, but also when a logic gate is subjected to inter- and intra-die variations (as discussed in Sect. 2.2.3). Because of the exponential sensitivities of variations in sub- or near-threshold operation, it is of the utmost importance to use *variation-resilient* circuit topologies.

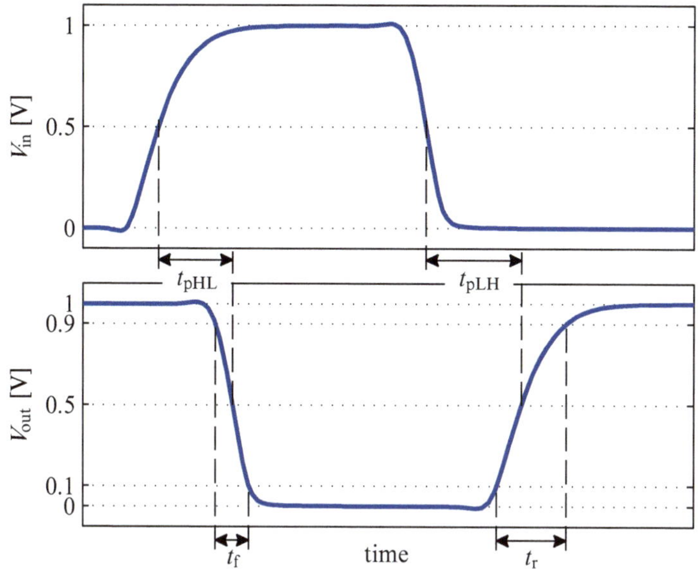

Fig. 3.3 Important transient characteristics of an inverter: definition of the delays

A third consideration for a circuit topology is the required *area* to implement a logic gate function. This concerns the area necessary to optimally size a logic gate for ultra-low-voltage operation. In general, larger area leads to higher capacitances, which deteriorate the operating speed and energy consumption. Furthermore, silicon area is proportional to cost, so if a certain circuit topology requires more area to implement a logic function, the cost of the total system will be higher. The area will be expressed as the equivalent amount of minimal transistors.

The earlier introduced *leakage power* and *dynamic energy* consumptions are of course essential characteristics in the comparison of circuit topologies, as they are crucial parameters throughout this entire work.

There are various ways to optimally size a standard CMOS logic gate for ultra-low-voltage operation. As could be seen in Fig. 2.11, a standard CMOS inverter which is regular-sized for operation at nominal supply voltage suffers severely from variability at ultra-low supply values. Dedicated sub-threshold sizing can counter this partly. To enable ultra-low-voltage operation for a standard CMOS inverter, the nMOS and pMOS transistor should be carefully balanced so that the noise margin is maximized [3]. The highest priority for optimal sizing is given to balanced noise margins in this work, since an imbalance results in a deteriorated nominal functionality at ultra-low supply voltages. If the nominal behavior is already skewed, the behavior under variations will emphasize this imbalance, especially in the snfp and fnsp corners. As a result, the variation-resilience decreases, which is very undesirable. Therefore, optimizing performance by balancing or minimizing propagation delays comes only on the second place, after guaranteeing robustness.

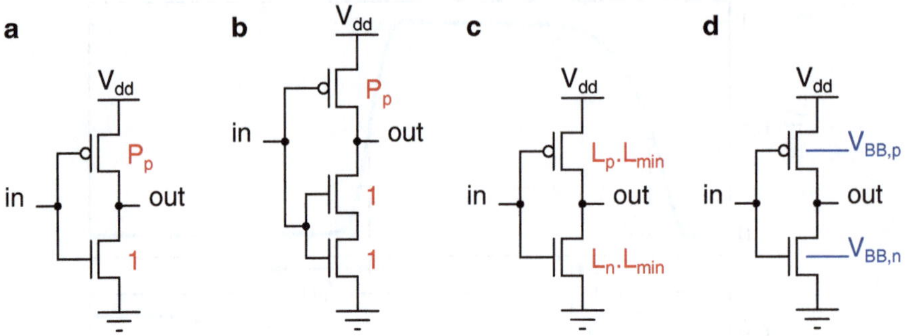

Fig. 3.4 Different possibilities to obtain optimal sizing for a standard CMOS inverter: (**a**) width sizing, (**b**) stacked nMOS, (**c**) length sizing and (**d**) body biasing

Naturally, it will be taken into account but it will not be the critical decisive factor. Moreover, many measures to improve performance can be taken on architectural level, which will be explained further in Chap. 4.

Different possibilities for this optimal sizing for a standard CMOS circuit topology will now be discussed: adjusting the width of the transistors, combining stacked transistors with adjusted width, adjusting the length of the transistors, and body biasing. The schematics visualizing these possibilities are provided in Fig. 3.4.

3.1.1.3 Width Sizing

The first option (Fig. 3.4a), optimal sizing through the width of the transistors, is the most commonly used one. The transistors' width is adjusted to balance the nMOS and pMOS transistor so that equal noise margins are obtained, as balanced noise margins allow to minimize the supply voltage at which the circuit is still functional. Since the on-current of nMOS transistors is significantly higher than the one of pMOS transistors for the same supply voltage, their width W_{nMOS} is kept minimal, so the factor W_{n} is equal to 1:

$$W_{\mathrm{nMOS}} = W_{\mathrm{n}} \cdot W_{\mathrm{min}} \tag{3.4}$$

$$W_{\mathrm{pMOS}} = W_{\mathrm{p}} \cdot W_{\mathrm{min}} \tag{3.5}$$

The pMOS width is then modified according to need. The relative width of the pMOS compared to the nMOS is expressed as P_{p}:

$$P_{\mathrm{p}} = \frac{W_{\mathrm{p}}}{W_{\mathrm{n}}} \tag{3.6}$$

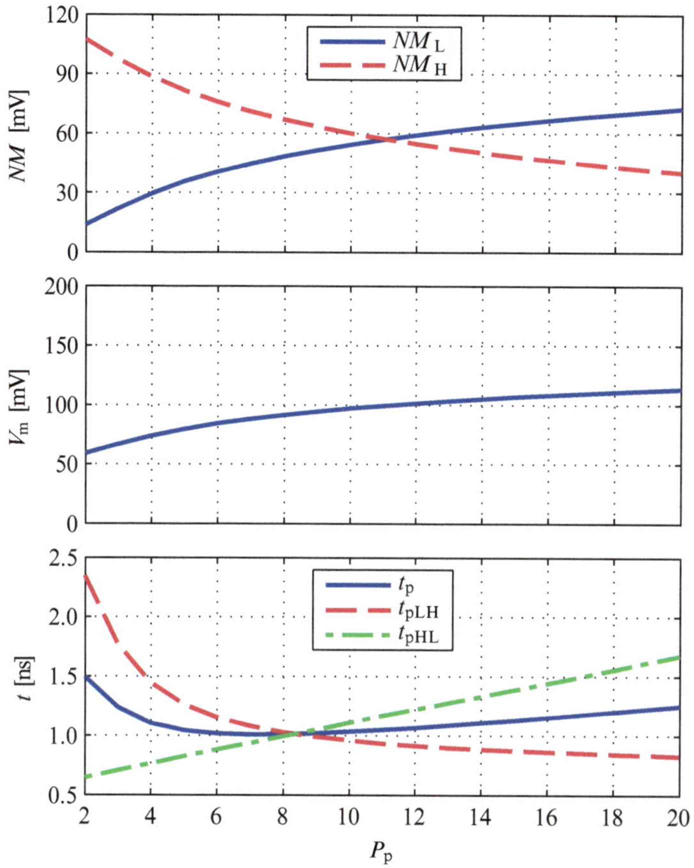

Fig. 3.5 Noise margins, switching threshold voltage and propagation delays of a standard CMOS inverter as function of the relative width P_p of the pMOS compared to the nMOS transistor (V_{dd} = 200 mV)

Figure 3.5 shows the various parameters which influence the optimal relative width P_p, obtained from simulations at a supply voltage of 200 mV. Equal noise margins are obtained at a P_p equal to 11.1. The value of P_p at which V_M is equal to $V_{dd}/2$ is 11.2. As could be expected, the optimal V_M occurs at a P_p value almost equal to the optimal value for equal noise margins, as they are closely related.

Another metric which is sometimes used for width sizing is obtaining equal propagation delays t_{pLH} and t_{pHL} or aiming for a minimal t_p. The former occurs at a P_p of 8.4, and the latter at 7. As can be seen, the required pMOS width is smaller to obtain minimum overall propagation delay. The reasoning behind this is that, while widening the pMOS improves the t_{pLH} of the inverter by increasing the charging current, it also degrades the t_{pHL} by causing a larger parasitic capacitance [31], as can be seen in Fig. 3.5.

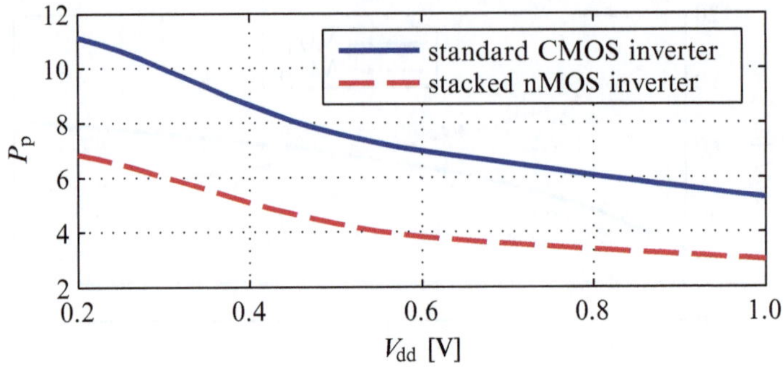

Fig. 3.6 Optimal P_p for different inverter implementations as function of V_{dd}

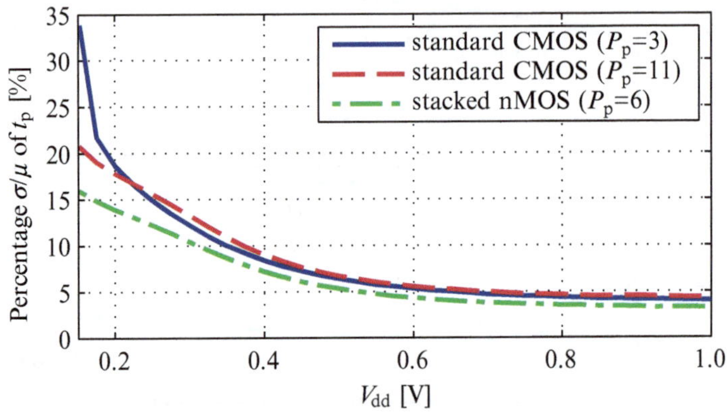

Fig. 3.7 Variation of t_p as function of V_{dd} for different inverter implementations

As explained above, priority is given to the sizing that acquires equal noise margins. Therefore, the pMOS width will be sized with a P_p of 11. This excessive sizing is a direct consequence of operating in the ultra-low-voltage region. To illustrate the influence of the supply voltage, Fig. 3.6 provides the optimal P_p as function of V_{dd} for the standard CMOS inverter. As can be seen, the required relative width increases significantly when lowering V_{dd}. Consequently, stacked pMOS transistors in for example a NOR gate require even more excessive sizes, which leads to large area and capacitance.

Nevertheless, this sizing is necessary to achieve the essential variation-resilience. Figure 3.7 proves that dedicated sizing of the standard CMOS inverter indeed counters the variation sensitivity (visualized earlier in Fig. 2.11) partly. Adequately sizing the inverter ($P_p = 11$) clearly lowers the variation of propagation delay in the ultra-low-voltage region compared to a regular-sized standard CMOS inverter ($P_p = 3$).

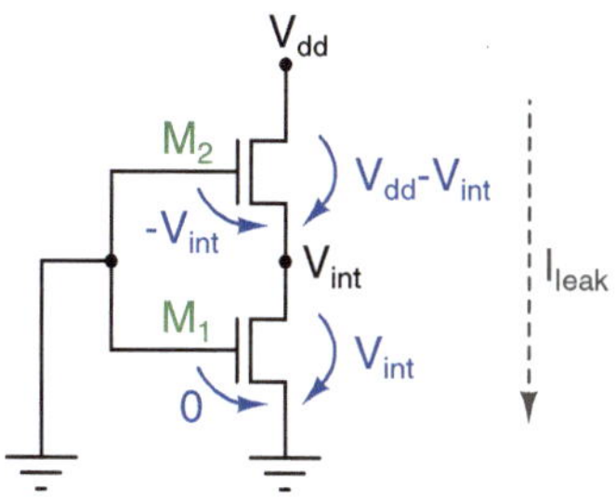

Fig. 3.8 Schematic of two stacked nMOS transistors, visualizing the different leakage reduction mechanisms

3.1.1.4 Transistor Stacking

A solution to this excessive pMOS sizing is to employ *transistor stacking*. Transistor stacking is a leakage reduction technique, based on the fact that two *off*-devices have significantly less leakage than a single *off*-device [26]. Figure 3.8 shows the schematic of two stacked nMOS transistors. Leakage is reduced by four different mechanisms [13]. They are all linked to the intermediate voltage V_{int} which is lower than V_{dd} and higher than V_{ss}:

1. The leakage current through transistor M_2 is reduced due to the negative gate-to-source voltage: $V_{\text{gs}} = -V_{\text{int}}$ (see Eq. (2.5)).
2. The body effect in M_2 increases $V_{\text{T,M}_2}$ due to the positive source-to-bulk voltage: $V_{\text{sb}} = V_{\text{int}}$ (see Eq. (2.9) and Fig. 2.4). M_2 is thus reverse body biased.
3. The DIBL effect in M_2 increases $V_{\text{T,M}_2}$ due to the reduced drain-to-source voltage: $V_{\text{ds}} = V_{\text{dd}} - V_{\text{int}}$ (see Fig. 2.6).
4. The DIBL effect in transistor M_1 increases $V_{\text{T,M}_1}$ due to the reduced drain-to-source-voltage: $V_{\text{ds}} = V_{\text{int}}$.

Therefore, transistor stacking is a very effective way of reducing leakage.

Stacking the nMOS transistor not only reduces the leakage current I_{off}, but its on-current I_{on} as well. Figure 3.9a shows the effect stacking has on the currents. As a result of the decreased $I_{\text{on,nMOS}}$ in a standard CMOS inverter with nMOS stacking, the pMOS sizing can be relaxed without degrading the noise margin. This is visualized in Fig. 3.9b, where the left axis shows the optimal P_{p} as function of the number of stacked nMOS transistors. The right axis shows that the overall noise margin remains balanced when this optimal P_{p} is used. The effect of stacking on the nMOS currents and thus on the pMOS sizing reduces with the amount of stacked transistors (as visible in Fig. 3.9). Therefore, it is optimal to stack the nMOS transistor twice, resulting in a relative pMOS width of 6.8. Figure 3.4b shows the schematic of such a stacked nMOS inverter, while Fig. 3.6 illustrates the relaxing effect this stacking has on the optimal P_{p} when sweeping the supply voltage.

Figure 3.10 shows the different characteristics which influence the optimal sizing for the stacked nMOS inverter. As already mentioned, optimal sizing for noise margins results in a P_{p} of 6.8. For V_{M}, the optimal P_{p} is 7.0. Equal propagation delays are achieved at a P_{p} of 5.0, while minimal t_{p} occurs at a P_{p} of 6.0.

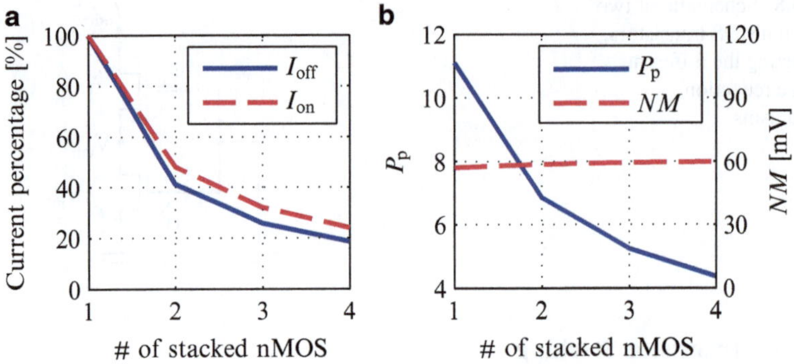

Fig. 3.9 For $V_{dd} = 200\,\text{mV}$: (**a**) Current percentage of stacked nMOS transistors relative to a single nMOS, as function of the amount of stacked nMOS transistors. (**b**) Relative width of pMOS P_p (left axis) as function of the number of stacked nMOS transistors in the inverter, in order to reach a maximal and balanced noise margin (right axis)

Consequently, the pMOS transistor for a stacked nMOS inverter has been chosen to be sized with a P_p of 6, to balance the different characteristics optimally. The total equivalent area is then $6 + 1 + 1 = 8$ for a stacked nMOS inverter, compared to $11 + 1 = 12$ for a standard CMOS inverter, which leads to a total area reduction of 33 %.

Stacking not only allows relaxed pMOS sizing, but decreases the leakage through the nMOS transistor as well, which reduces the static power consumption with 58.7 % compared to the standard CMOS inverter (at $V_{dd} = 200\,\text{mV}$). The stacked nMOS inverter thus has an area and a leakage power reduction, at the penalty of an increased nominal propagation delay of 23.5 % compared to the standard CMOS inverter.

However, the stacked nMOS inverter has a positive effect on delay variations in comparison to the standard CMOS inverter. Figure 3.7 has proven that adequately sizing the standard CMOS inverter to sub-threshold restrictions lowers the variation of t_p at ultra-low supply voltages compared to a regular-sized standard CMOS inverter. However, it also shows that using a stacked nMOS inverter further decreases the delay variation. To summarize, introducing nMOS stacking increases the nominal propagation delay slightly, but it also reduces the percentage variation of the delay with 3.9 % at 200 mV, as visible in Fig. 3.7. Due to the variation-resilience of the stacked nMOS inverter, lower design margins have to be introduced to cope with timing variations compared to conventional standard CMOS inverters.

3.1.1.5 Length Sizing

Up to now, the weaker pMOS transistor was strengthened by increasing its width, so as to obtain equal drive strengths of both transistors and consequently equal noise margins. Another method to counter the stronger nMOS transistor would be

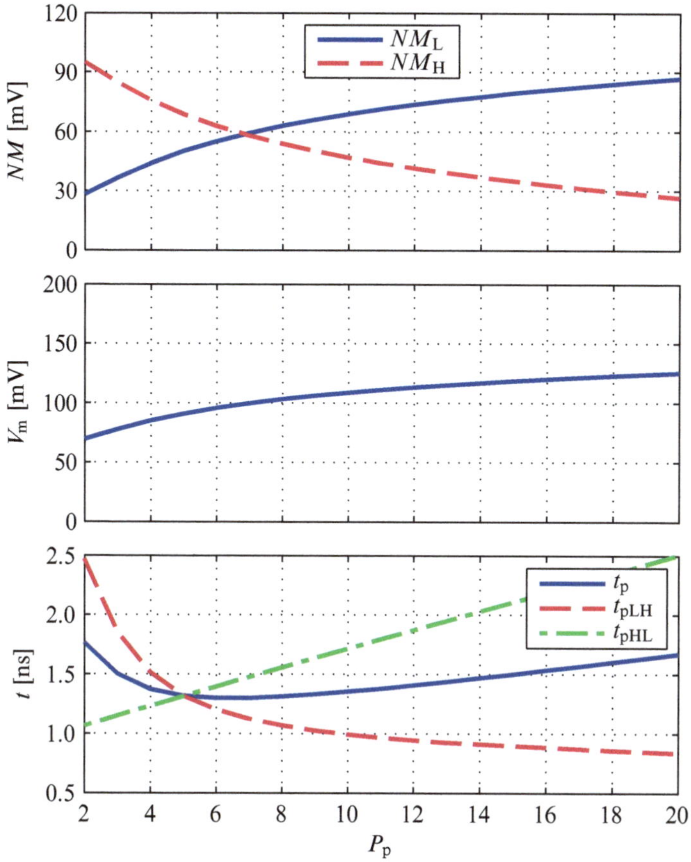

Fig. 3.10 Noise margins, switching threshold voltage and propagation delays of a standard CMOS inverter with stacked nMOS transistors as function of the relative width P_p ($V_\mathrm{dd} = 200\,\mathrm{mV}$)

to weaken it by increasing its length L_nMOS while keeping its width W_nMOS and the width and length of the pMOS transistor minimal. This should result in a decreased current in all operating regions of the transistor, as visible in the current equations of Sect. 2.1.1. Increasing the length is therefore sometimes used in digital circuits to reduce the leakage or to limit the on-current of a transistor. The resulting schematic of a standard CMOS inverter with nMOS length sizing is shown in Fig. 3.4c, where L_p would be kept equal to 1 and L_n would be sized according to the required needs.

However, in the 90 nm CMOS technology at hand, the transistor models demonstrate strange behavior when adapting the length of an nMOS transistor. This can be seen in Fig. 3.11, where I_off and I_on are normalized to an nMOS of minimal size and plotted as a function of L_nMOS. Both currents increase with higher length, instead of the expected decrease. Table 3.1 provides a comparison of the impact of

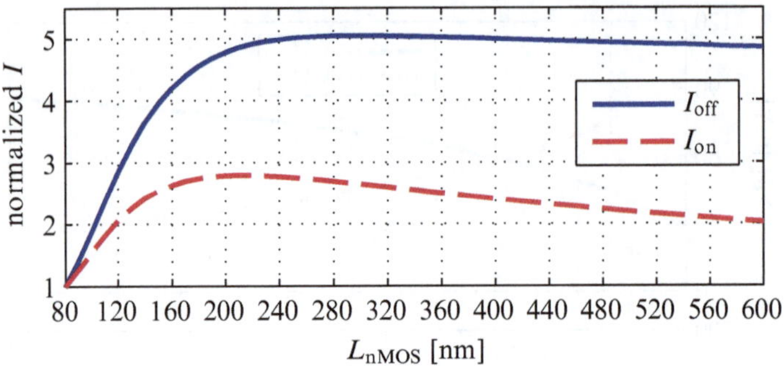

Fig. 3.11 Normalized currents $I_{\text{on,nMOS}}$ and $I_{\text{off,nMOS}}$ as function of L_{nMOS} ($W_{\text{nMOS}} = W_{\text{min}}$ and $V_{\text{dd}} = 200\,\text{mV}$)

Table 3.1 Comparison of normalized I_{on} and I_{off} of nMOS transistors with different sizings

nMOS transistor	Normalized I_{off}	Normalized I_{on}
W_{min}, L_{min}, no stacking	1.00	1.00
W_{min}, $2{\cdot}L_{\text{min}}$, no stacking	4.20	2.63
W_{min}, L_{min}, stacked twice	0.41	0.48

nMOS sizing on its currents. According to the simulations, doubling the length of the nMOS transistor results in an increase of a factor 4.20 in leakage current, while nMOS stacking reduces I_{off} with a factor of 0.41.

As already discussed in Sect. 2.3.3, the accuracy of transistor models in the weak inversion region is not always reliable. This might be a model artifact, but ignoring it to instead rely on intuitive transistor behavior severely complicates circuit simulations. Furthermore, the different mechanisms influencing transistor stacking actually result in a larger leakage reduction than doubling the channel length of the transistor [13]. Moreover, in modern deep sub-micron devices the Reverse Short-Channel Effect (RSCE) may reduce the threshold voltage of the transistor for longer channels, resulting in a less effective leakage reduction. In general, there is a high sensitivity of the current as function of the transistor's length to process and technology parameters. Hence, sizing of standard CMOS logic gates through adjusting the length of the transistors is strongly technology-dependent [4].

3.1.1.6 Body Biasing

A fourth method to balance the on-currents of the nMOS and pMOS transistor is to employ the body effect (as explained in Sect. 2.1.2.2) to make the transistors weaker or stronger. Reverse Body Biasing (RBB) increases V_{T} to obtain less leakage at the cost of decreased performance. Forward Body Biasing (FBB), on the other hand, reduces V_{T} to increase performance at the cost of higher leakage. RBB is achieved by reducing $V_{\text{BB,n}}$ below the ground rail for nMOS transistors, and by increasing $V_{\text{BB,p}}$ above the supply rail for pMOS transistors. Equivalently, increasing $V_{\text{BB,n}}$ with respect to V_{ss} and reducing $V_{\text{BB,p}}$ with respect to V_{dd} results in FBB.

This sizing method is visualized in Fig. 3.4d for the standard CMOS inverter. Here, RBB through decreasing $V_{BB,n}$ could be used to increase the threshold voltage of the nMOS transistor, up to the point where its I_{on} matches that of a minimal pMOS. Oppositely, FBB through decreasing $V_{BB,p}$ could make the pMOS transistor stronger by reducing its threshold voltage.

However, introducing body biasing has some consequences. Firstly, additional power supply rails to distribute the body biasing voltages, as well as a triple well technology are required. Charge pump circuits are needed to generate the additional supplies. This results in area and energy overhead. Secondly, to compensate for inter-die variability, body biasing can be employed but this requires calibration after fabrication. Each individual die then needs to be calibrated during initial measurements. Thirdly, the impact of body biasing reduces for short-channel devices, thereby affecting the scalability of this method. Fourthly, the body biasing voltages are limited by latch-up on one side and electrical breakdown on the other side. Especially in advanced nanometer technologies where the body effect coefficient γ is reduced, these limits can restrict the effectiveness of body biasing.

Both body biasing techniques have their own separate issues as well. RBB becomes less effective for leakage reduction at shorter channel lengths [41]. RBB increases the sensitivity to process variations, e.g. it worsens the V_T variations across a die [28]. FBB reduces the sensitivity to process variations, but suffers severely from temperature dependencies [27].

3.1.1.7 Sizing Conclusion

Because of the aforementioned restrictions of length sizing and body biasing of standard CMOS logic, these two options will be discarded in the remainder of this circuit topology comparison. As explained in Chap. 1, this book is focused on ultra-low-voltage circuit design in bulk CMOS technologies. This conclusion is therefore only valid for these type of technologies and could be different in other technologies, e.g. SOI.

3.1.1.8 Literature

There are many ultra-low-voltage publications which utilize standard CMOS logic, as these gates are readily available in standard cell libraries. In some cases, regular standard cells are used which have been resimulated at low target voltages to check their functionality at such supplies, the unfunctional cells were then discarded. For example, standard CMOS gates with large stacks were often avoided, e.g. in [8, 15, 16, 19, 20, 23, 24, 30]. In most cases, recharacterization of the standard cell library at low supply voltages has been carried out [1, 7, 17, 25, 37, 48].

Sizing for sub-threshold operation has been done with both width sizing and length sizing, but the former method has been much more often used than the latter. Hanson et al. [11] and Bol et al. [6] have suggested to increase the channel length

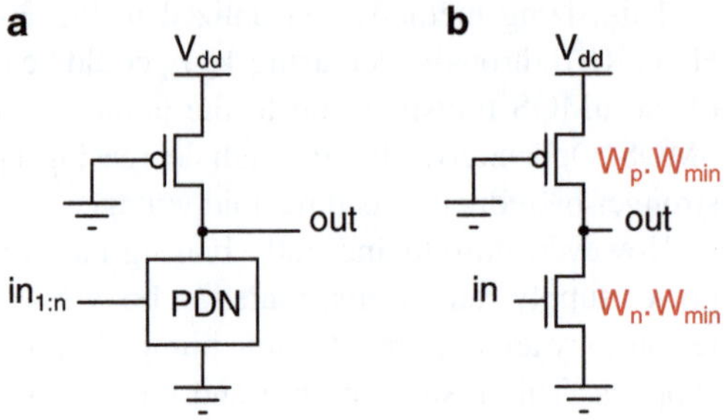

Fig. 3.12 (**a**) Generic implementation of an *n*-input pseudo-nMOS logic gate. (**b**) Schematic of a pseudo-nMOS inverter

to improve the transistor's sub-threshold behavior. In measured ultra-low-voltage designs, length upsizing has been employed by for instance [9, 24].

Body biasing has been extensively used to compensate for variations after manufacturing of sub- and near-threshold designs, e.g. in [10, 12, 14, 18, 21, 44]. However, as shown in Sect. 1.5, the designs using body biasing do not outperform the other designs.

3.1.2 *Pseudo-nMOS Logic*

3.1.2.1 Concept

Figure 3.12a shows the generic implementation of an *n*-input *pseudo-nMOS* logic gate. The PDN is identical to the PDN of a standard CMOS logic gate, but the PUN has been replaced by a single pMOS transistor that is grounded so that it acts as a current source. Hence, the PDN realizes the logic function, while the pMOS transistor functions as load. When the PDN is off, the pMOS load pulls the output to '1'. When the PDN turns on, it fights the load. Therefore, the pMOS load must be weak enough so that the output pulls down to an acceptable '0' level. In order for this logic gate to work correctly, the pMOS sizing is thus critical.

Pseudo-nMOS logic is a form of so-called *ratioed* logic. In general, ratioed circuits depend on device sizing to produce acceptable output levels. In ratioless logic, on the other hand, the output levels do not depend on the sizing of the devices. The other topologies discussed in this chapter are all ratioless circuits.

3.1.2.2 Ultra-Low-Voltage Operation

Ratioed logic reduces the number of transistors to implement a given logic function with respect to standard CMOS logic: $N + 1$ transistors are required instead of $2N$. The advantage of ratioed logic is the decreased number of devices and the smaller area. However, ratioed logic introduces several disadvantages.

Figure 3.12b shows the schematic of a pseudo-nMOS inverter. When the output is pulled high, the operation is the same as for a standard CMOS inverter. When the output is pulled low, the nMOS transistor is turned on while the pMOS load also conducts current. This has two important consequences. Firstly, the nominal low output voltage is higher than V_{ss}, resulting in a decreased low noise margin NM_L. Secondly, the inverter has a large static power dissipation due to the direct path from the supply to the ground in the low output state.

The area reduction thus comes at the cost of decreased robustness and static leakage. In fact, the sizing of the pseudo-nMOS logic gate results in a trade-off between noise margin, power dissipation, and delay [31]. The first two parameters get worse as the pMOS size increases. On the other hand, a smaller pMOS results in a lower rise time. Since robustness and leakage are of primary concerns for ultra-low-voltage operation, the pMOS transistor will be sized minimally in this implementation. The nMOS transistor is then sized in order to obtain equal noise margins. This results in a W_p of 1 and a W_n of 6 at a 200 mV supply for the 90 nm CMOS technology at hand.

Figure 3.13 visualizes the most important drawback of pseudo-nMOS logic: its sensitivity to variations. In this figure, the VTCs of a pseudo-nMOS inverter ($W_p = 1$, $W_n = 6$) and a standard CMOS inverter ($P_p = 11$) are compared in different process corners at a supply of 200 mV. As can be seen, the standard CMOS inverter displays good behavior under inter-die variations when properly sized for ultra-low-voltage operation. The pseudo-nMOS inverter on the contrary, suffers severely from these inter-die variations, even though it is properly sized for nominal operation. This makes pseudo-nMOS logic unusable in the ultra-low-voltage region.

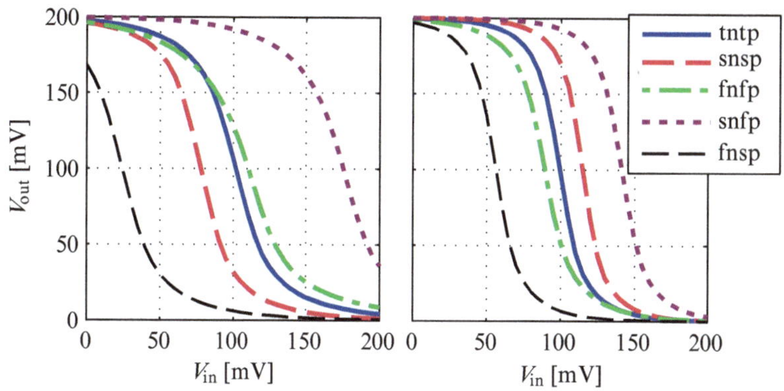

Fig. 3.13 VTC in process corners of a pseudo-nMOS inverter (*left*) in comparison with a standard CMOS inverter (*right*) at $V_{dd} = 200$ mV

3.1.2.3 Literature

Pseudo-nMOS logic has been introduced by Soeleman and Roy in 1999 as a favorable circuit topology for use in the sub-threshold region [35]. Listed advantages were the reduced area and the improved performance due to the reduction of the load capacitance, in comparison to standard CMOS logic. This was one of the first groups who performed research on sub-threshold logic and ultra-low-voltage operation in general, and they published a few papers on simulation results of sub-threshold pseudo-nMOS logic until 2001. However, it has not been adopted by other groups, because of the unacceptably high sensitivity to variations.

3.1.3 Pass Transistor Logic

3.1.3.1 Concept

The previously discussed logic families only allow inputs to drive the gate terminal of a transistor. *Pass transistor* logic is a circuit topology which not only allows inputs to drive gate terminals, but source/drain terminals of transistors as well. Fundamentally, transistors are used as switches, as shown in Fig. 3.14. A single pass transistor can be realized with an nMOS or a pMOS transistor. Logic gates can easily be constructed with pass transistor logic, e.g. Fig. 3.14c presents the schematic of a NOR gate implemented with nMOS pass transistor logic. Compared to a standard CMOS implementation of a NOR gate, pass transistor logic requires much less transistors. Historically, this is the main motivation behind the use of pass transistor logic.

Note that pass transistor logic is still *static* logic, as are the previously discussed circuit topologies. The outputs of static logic are always connected to either V_{dd} or V_{ss} through a low resistive path, which is advantageous for noise resilience [31]. It is clear that for example in the NOR gate depicted in Fig. 3.14c always one of the

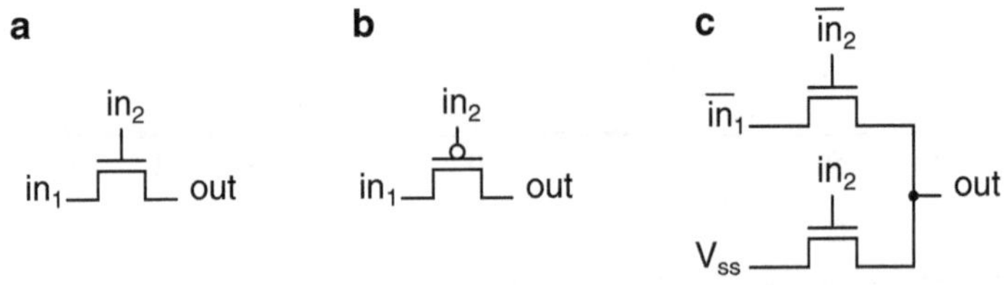

Fig. 3.14 Schematics of pass transistor logic: (**a**) nMOS switch, (**b**) pMOS switch and (**c**) nMOS NOR gate

pass transistors will be conducting, ensuring the static property of pass transistor logic. Section 3.1.5.3 will discuss *dynamic* logic, which relies on temporary storage on the capacitance of a high impedance node.

Unfortunately, pass transistor logic always suffers from signal loss: nMOS transistors pass a strong '0' but a weak '1', i.e. a V_T loss will occur on the logic high level. Equivalently, pMOS transistors pass a strong '1' but a weak '0', i.e. a V_T loss will occur on the logic low level. Because of this inherent V_T loss, pass transistor gates cannot be cascaded by connecting the output of a pass transistor to the gate input of a subsequent pass transistor.

3.1.3.2 Ultra-Low-Voltage Operation

This inherent V_T loss makes pass transistor logic unsuitable for operation at ultra-low supply voltages. The voltage drop could be solved by pulling the output to the supply rails, but this requires additional circuitry after every logic gate. Adding an inverter could for example ensure this level restoration. However, the extra transistors added for the additional level restoring circuitry compromise the benefit of pass transistor logic, which was the low transistor count. A more elegant solution to the voltage drop will be proposed as the next circuit topology.

3.1.3.3 Literature

One of the first differential implementations of pass transistor logic is described in [47]. A 16-bit multiplier is constructed with Complementary Pass transistor Logic (CPL). CPL consists of differential inputs and outputs, an nMOS-only pass transistor logic network and standard CMOS output inverters. Basically, logic gates are constructed with differential inputs and nMOS pass transistors. The main reason why [47] used CPL was to achieve high speed due to lower input capacitance and higher logic functionality. The published circuits utilizing CPL were functioning at nominal supply.

The only pass transistor based family which was designed to be used at ultra-low supply voltages has been proposed by the Berkeley Wireless Research Center in 2007 [2]. The so-called Sense Amplifier-based Pass Transistor Logic (SAPTL) consists of three major components. Firstly, there is a pass transistor tree, called the stack, which computes the desired logic function. An inverter drives the root node of the stack and injects signals into the stack. At the output of the stack, a sense amplifier is used to recover both voltage swing and performance. The drivers and sense amplifiers thus provide gain to the circuit. Since the pass transistor stack has no V_{dd} or V_{ss} connections, the only leakage paths appear in the gain circuits. SAPTL can operate synchronously using a clock, or asynchronously using additional hand-shaking circuitry. The authors claim that the low leakage and the low energy consumption are the main advantages of SAPTL. However, the supply voltage that can be used in SAPTL is limited by the input voltage difference that the sense amplifiers can sense. To decrease the input swing of the sense amplifier, its design becomes more difficult and its area or energy consumption will probably increase.

3.1.4 Transmission Gate Logic

3.1.4.1 Concept

The main disadvantage of pass transistor logic is the V_T loss at one of the signal levels. This can be solved by using the complementary properties of nMOS and pMOS transistors: instead of placing a single transistor to pass a signal, two complementary transistors could be placed. This is called a *transmission gate*, and is visualized in Fig. 3.15a. While switching, current will flow through the parallel combination of the nMOS and pMOS transistor. The nMOS passes a strong '0', while the pMOS passes a strong '1', thereby eliminating the V_T loss on both logic levels.

When this technique is used to implement logic gates, it is called Transmission Gate (TG) logic. Figure 3.15b shows a NOR gate implemented with TG logic. Compared to pass transistor logic (recall Fig. 3.14c), TG logic requires double the amount of transistors, but it eliminates the problematic voltage drop. TG logic is commonly built using equal-sized minimal nMOS and pMOS transistors. Boosting the size of the pMOS, as in standard CMOS logic, only slightly improves its effective resistance while significantly increasing the capacitance [46]. As opposed to standard CMOS logic, there is no need for transistor balancing through sizing in TG logic since there is always an nMOS and a pMOS included in a conducting path. Compared to a standard CMOS NOR gate, the required area is therefore much lower. Hence, TG logic is still attractive from an area point of view, despite the transistor doubling compared to pass transistor logic.

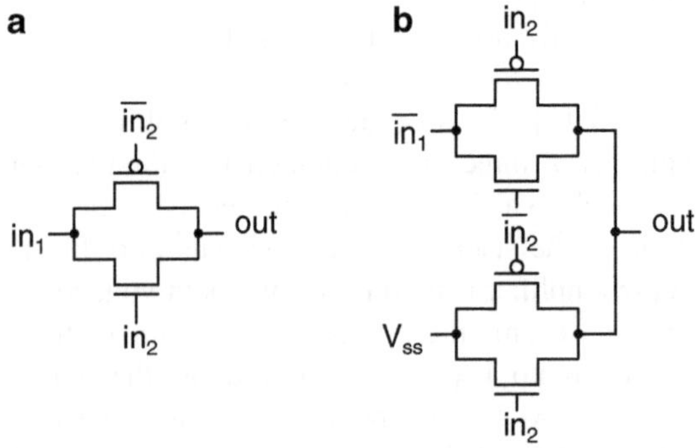

Fig. 3.15 Schematics of transmission gate logic: (**a**) transmission gate and (**b**) TG NOR gate

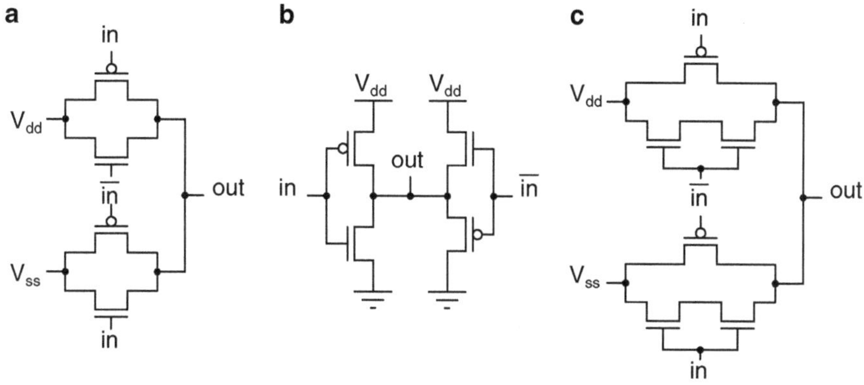

Fig. 3.16 Schematics of an inverter: (**a**) in TG logic, (**b**) alternative representation in TG logic, (**c**) in stacked nMOS TG logic

Note that TG logic requires complementary input signals, as can be seen in Fig. 3.15. The required extra wires increase routing complexity, as opposed to standard CMOS or pseudo-nMOS logic.

3.1.4.2 Ultra-Low-Voltage Operation

This section will provide an in-depth analysis of TG logic, and a detailed comparison to standard CMOS logic will be performed.

One of the attractive properties of TG logic in ultra-low-voltage operation is that it suffers less from reliability issues due to inter-die variations compared to standard CMOS logic. An intuitive explanation will first be discussed, and will afterwards be followed by supporting simulation results. Figure 3.16a shows the schematic of an inverter implemented in TG logic, while Fig. 3.16b provides an alternative representation of the same TG inverter. This alternative representation shows that the TG inverter is in fact a standard CMOS inverter, extended with an 'inverse' standard CMOS inverter. The inverse inverter has the complementary input signal of the regular inverter, and has an nMOS in its PUN and a pMOS in its PDN. The process corners which are most problematic from a functionality perspective are the fnsp and snfp corners where the speed difference of the transistors is largest. Exactly for these corners, this inverse inverter aids significantly, since there are always both an nMOS and a pMOS in parallel that can compensate each other's weaknesses.

Before evaluating process corner simulation results, the exact sizing of TG logic in the 90 nm technology at hand must be discussed. The TG logic implementation of Fig. 3.16a with a single nMOS and pMOS transistor in each transmission gate actually poses problems. These problems are related to the I_{on}/I_{off} ratios discussed in Sect. 2.2.2. In this technology, $I_{off,nMOS}$ is only 21.7 times lower than $I_{on,pMOS}$ at $V_{dd} = 200\,mV$. Extensive MC simulations of the $I_{on,pMOS}/I_{off,nMOS}$ ratio resulted

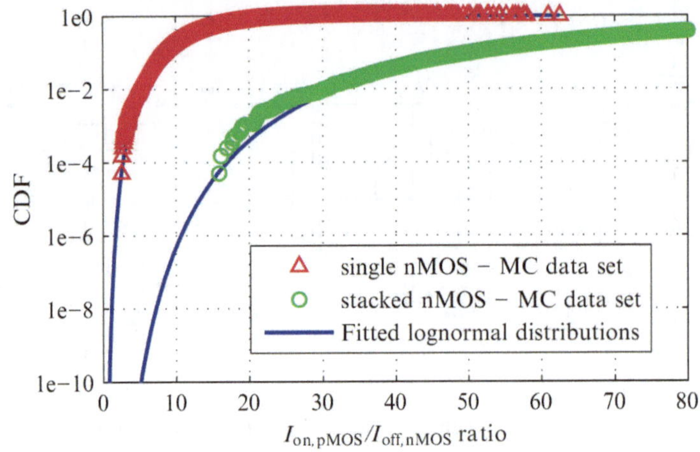

Fig. 3.17 Critical tail of the CDF of the $I_{on,pMOS}/I_{off,nMOS}$ ratio obtained with MC simulations. Both the single nMOS and the stacked nMOS implementations are fitted with lognormal distributions ($V_{dd} = 200\,\text{mV}$)

in a CDF of which the critical tail at low current ratios is shown in Fig. 3.17. Nominally the current ratio is already very low, but taking into account 6σ intra-die variations the worst-case ratio becomes insufficient, as can be seen from the fitted lognormal distribution on the lower end tail of the ratio with a single nMOS. An important point to make is that the $I_{on,pMOS}/I_{off,nMOS}$ ratio is much smaller than the $I_{on,nMOS}/I_{off,pMOS}$ ratio, as already shown in Fig. 2.8.

To improve this problematically low $I_{on,pMOS}/I_{off,nMOS}$ ratio, the nMOS transistor is stacked (as shown in Fig. 3.16c). This results in a decreased $I_{off,nMOS}$ (see Table 3.1). Stacking the nMOS thereby mitigates the current ratio problems, since it increases the $I_{on,pMOS}/I_{off,nMOS}$ ratio while the complementary current ratio remains sufficiently high. Moreover, using stacked nMOS transistors results in a significantly higher worst-case current ratio than using a single nMOS (Fig. 3.17). Without nMOS stacking, there is a large difference between the rise and fall time of a TG. The rise time is the critical timing specification because it is dominated by the weak pMOS. Stacking the nMOS transistor of TG logic results in a more balanced rise and fall time and thus has a negligible effect on the overall speed of the logic gate. To conclude, the increased robustness and the reduced leakage outweigh the slight speed degradation cause by nMOS stacking.

Note that nMOS stacking is here used to decrease $I_{off,nMOS}$, while in the case of standard CMOS, the main reason for the use of nMOS stacking is the reduced $I_{on,nMOS}$.

To evaluate the aforementioned inter-die variation-resilience of TG logic, Fig. 3.18 provides the simulated VTCs in process corners of a stacked nMOS TG inverter ($W_p = 1$ and $W_n = 2$ in Fig. 3.16c) and a standard CMOS

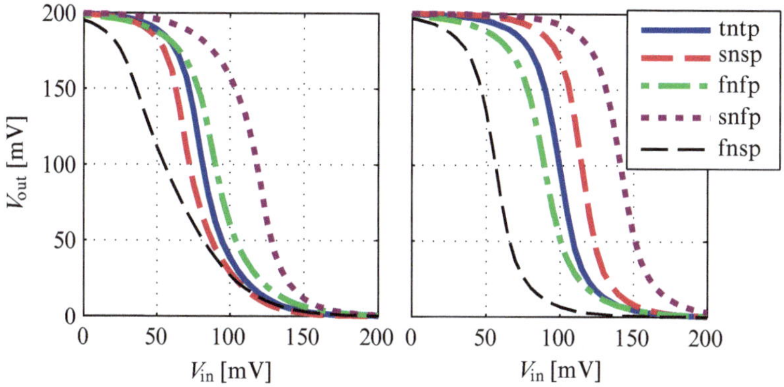

Fig. 3.18 VTC in process corners of a stacked nMOS TG inverter (*left*) in comparison with a standard CMOS inverter (*right*) at $V_{dd} = 200\,\text{mV}$

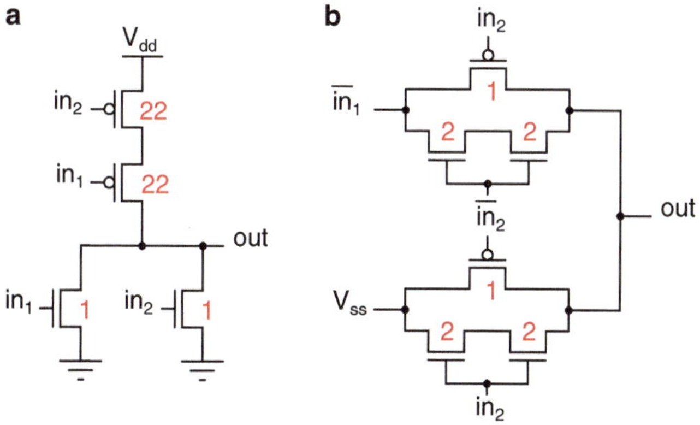

Fig. 3.19 Schematics of a NOR gate: (**a**) in standard CMOS logic and (**b**) in stacked nMOS TG logic

inverter ($P_p = 11$). It can be seen that the TG inverter has less spread over the different process corners than the standard CMOS inverter. The higher inter-die variation-resilience of TG logic arises from the inclusion of both nMOS and pMOS transistors in each conducting path. With stacked nMOS TG logic, functionality is thus ensured under all possible inter-die variations.

Until now, the analysis on TG logic has been on an inverter. However, using this topology for logic gates has far more interesting benefits, of which the area is only one. In the following analysis, standard CMOS logic with pMOS width upsizing (abbreviated to CMOS) and TG logic extended with nMOS stacking (abbreviated to TG) are compared on various logic gate characteristics. The analysis will be performed on a NOR gate because it is an elementary logic function and a difficult gate in standard CMOS logic since it requires pMOS stacking. Stacked pMOS transistors in NOR gates require excessive sizes, as can be seen in Fig. 3.19a.

The sizing of the TG NOR is shown in Fig. 3.19b. The pMOS transistors are sized minimally, and the nMOS transistors are stacked and have a width of $2 \cdot W_{min}$. At first sight, this seems counterintuitive because increasing the width of a transistor normally increases its I_{on} and I_{off}. However, in this 90 nm CMOS technology, I_{off} and I_{on} of stacked nMOS transistors with $2 \cdot W_{min}$ reduce with 54 % and 27 % compared to minimal-sized stacked nMOS transistors, respectively. This is due to the *Inverse Narrow Width Effect* (INWE) (also called Reverse Narrow Channel Effect) of which the impact in the sub-threshold region has been discussed in [49]. INWE only has an impact for transistor widths that approach the minimum width: it effectively reduces the threshold voltage for very narrow transistor widths. Therefore, slightly increasing the nMOS width is beneficial to further reduce its leakage. Note that INWE is only present in the 90 nm and not in the 40 nm CMOS technology used in this work. As a result, stacked nMOS transistors in TG logic of the 40 nm prototypes are sized minimally, as will be seen in Sect. 3.4. To summarize, the sizing of TG logic is relaxed considerably compared to CMOS logic, e.g. the area of the CMOS NOR gate is 4.6 times bigger than the TG NOR.

In the analysis, the NOR gate is subjected to inter- and intra-die variations. Due to the exponential sensitivity to variations, it is of the utmost importance to design variation-resilient sub-threshold circuits.

Because of the small supply voltage swing, an important characteristic in ultra-low-voltage design is the output signal loss of logic gates. Too much signal loss can cause the subsequent gate to wrongly interpret the logic value. Signal losses can be overcome by regenerating the signal, e.g. through an inverter. For example in a datapath with a high logic depth, intermediate signals of cascaded logic gates can be regenerated to ensure correct output levels. However, the lower the amount of signal loss, the less frequently inverters need to be inserted to restore the signal levels to the supply rails.

Figure 3.20 compares the TG and CMOS NOR gates on the percentage signal loss their output has relative to the total supply swing, under inter-die variations. Only the worst-case corners are shown as a function of V_{dd}. In the case of signal loss on the logic low level, the logic gates perform worst in the snfp corner because of the weakened nMOS transistor versus the strengthened pMOS transistor. Respectively, at signal loss on the logic high level, this worst-case applies to the fnsp corner. Figure 3.20 shows that the signal loss aggravates when the supply voltage lowers and the circuits operate more in sub-threshold. It is clear that the TG NOR outperforms the CMOS NOR in signal loss on logic low level, and TG logic is also the better option in the case of signal loss on logic high level. The output swing degradation analysis is performed for intra-die variations as well, by carrying out extensive MC simulations for a 200 mV supply. Figure 3.21 demonstrates that the TG NOR performs significantly better under intra-die variations for signal loss on logic low level and comparably for logic high level.

Note that this signal loss story changes when cascading multiple gates in TG logic, as will be discussed profoundly on architectural level in Sect. 4.2.

Another essential characteristic is the variation of gate delay. As previously mentioned, intra-die variations have a very deteriorating influence on the variation

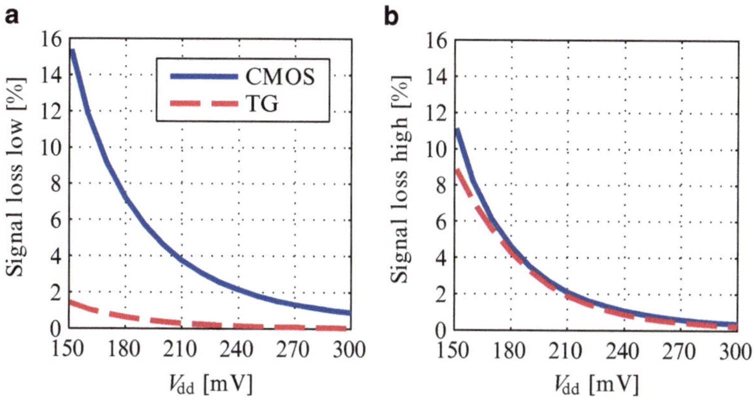

Fig. 3.20 Percentage signal loss for different NOR topologies in the worst-case corner: (**a**) snfp corner for logic low and (**b**) fnsp corner for logic high level, as function of V_{dd}

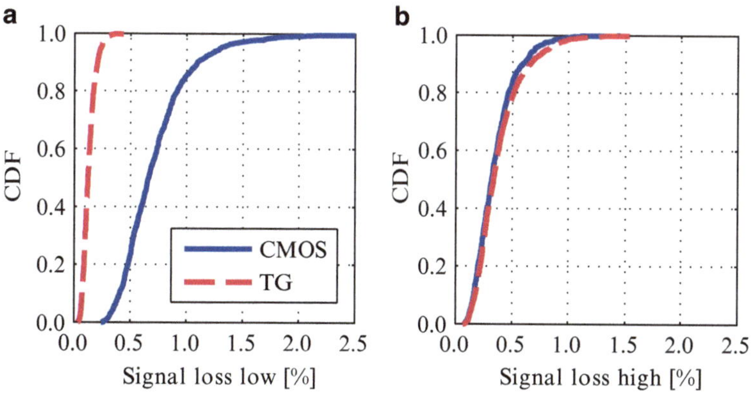

Fig. 3.21 CDF of the signal loss for different NOR topologies of (**a**) logic low and (**b**) logic high level for $V_{dd} = 200\,\text{mV}$, obtained with Monte Carlo (MC) simulations around the tntp corner

in delay. Therefore, Fig. 3.22 shows the variation of the propagation delay as function of V_{dd}. The TG NOR displays overall less delay variations than the CMOS NOR.

An additional, important benefit of TG logic is the fact that it does not have direct leakage paths from the supply to the ground. As such, a TG logic gate has the attractive property of an almost non-existing leakage power. Table 3.2 provides the leakage power figures of both NOR topologies. The leakage power of the CMOS NOR is a factor of more than 19 higher than the one of the TG NOR. These numbers take only the inherent leakage of the NOR gates into account, not the contribution to leakage of possible circuits required to regenerate intermediate signal levels, which will be examined in detail later on.

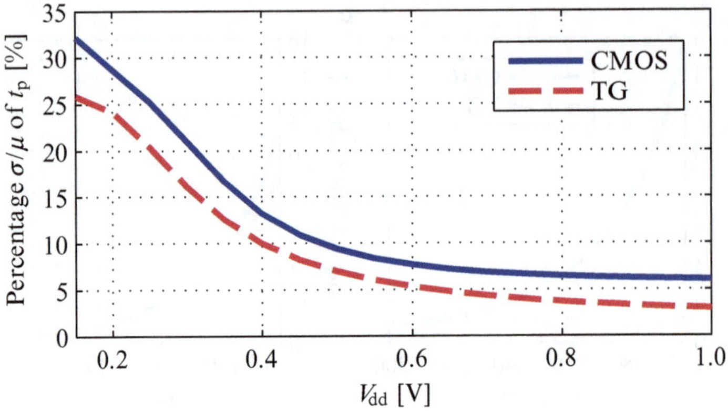

Fig. 3.22 Variation of the propagation delay for different NOR topologies as function of the supply, obtained with Monte Carlo (MC) simulations

Table 3.2 Comparison of leakage power for different NOR topologies ($V_{dd} = 200\,mV$)

	CMOS NOR	TG NOR
Leakage power	3125 pW	161 pW

To conclude this analysis, TG logic is a very attractive solution for ultra-low-voltage logic gates due to its higher variation-resilience and lower leakage than standard CMOS logic. This analysis has been performed on a NOR gate, because it is an elementary logic function. Important to note is that since all logic gates in TG logic have the same generic structure, the results for other TG logic gates will be very similar to the ones of the NOR. More information about this will be provided in Sect. 3.2.1.

3.1.4.3 Literature

In literature, transmission gates have been reported to be used in an ultra-low-voltage design once by another research group. Wang and Chandrakasan from MIT [42, 43] presented a sub-threshold FFT processor where transmission gate logic was used, but only for a few specific logic gates, e.g. a XOR and a MUX. This concerned regular transmission gate logic, so no transistor stacking was employed. However, to avoid sneak leakage paths and thus ensure functionality, they buffered all inputs and outputs to the transmission gate cells. This of course adds significantly to the resulting leakage and energy of inserting such a TG cell.

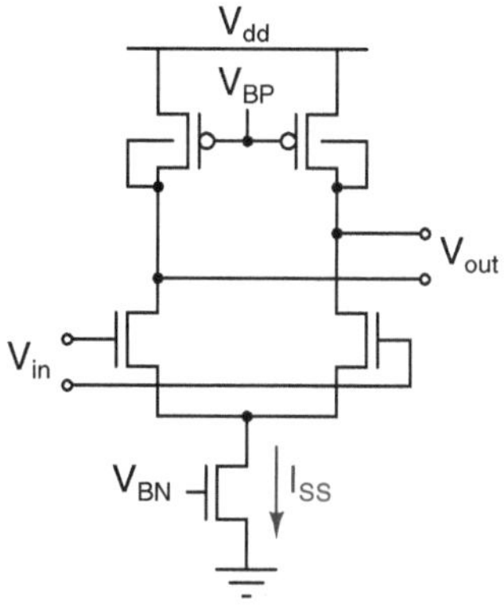

Fig. 3.23 Schematic of an STSCL inverter

3.1.5 Other Topologies

This section covers some other, less frequently used circuit topologies for ultra-low-voltage or ultra-low-energy operation.

3.1.5.1 Sub-Threshold Source-Coupled Logic

Sub-Threshold Source-Coupled Logic (STSCL) has been proposed by Tajalli and Leblebici from the Ecole Polytechnique Fédérale de Lausanne (EPFL) in Switzerland [39]. Figure 3.23 shows the schematic of an STSCL inverter. In an STSCL gate, the logic operation takes place mainly in the current domain to achieve a very high speed. The input source-coupled nMOS differential pair switches a constant current between two branches, based on the input logic levels. This differential pair can be expanded to a network of nMOS source-coupled pairs to implement more complex logic functions. The current is converted to an output voltage through the pMOS load transistors. The voltage swing at the output should be large enough to completely switch the current in the input transistors of the next stage. Hence, the load resistors should have a high enough resistivity. Minimal-sized pMOS transistors with shorted drain-substrate contacts are used as gate-controlled, highly resistive load devices. The bias current (through the nMOS transistor below) is usually kept at very low current levels.

Operating in sub-threshold regime, the circuit can be used in a very wide frequency range by adjusting the bias current without any need for resizing the devices. The power consumption of an STSCL gate depends on the tail bias current. Unlike standard CMOS circuits where there is no constant current dissipation, each STSCL gate consumes a certain amount of constant bias current. This current is charging or discharging the load capacitance, and thus directly translates into the speed of the output transition. The most interesting aspect of STSCL circuits is that both speed and power consumption can be adjusted linearly by altering the amount of bias current. Hence, this allows a wide range of operating frequencies. However, because of this static power consumption, STSCL logic is mostly power-efficient in circuits with high activity. Evidently, bias circuits are required to provide bias currents of both nMOS current source and the pMOS loads.

To maintain enough headroom for the current source, a minimum supply voltage of around $10 \cdot V_{th}$ is necessary [39]. Measurements of an 8-bit carry-save multiplier in a 0.18 μm CMOS technology [38] confirm that this theoretical value is approximately correct, since the multiplier is functional down to 300 mV. Hence, extremely low-supply operation is not possible with STSCL circuits. Unfortunately, no larger STSCL systems than this multiplier have been fabricated, making it difficult to assess the scalability of this type of logic. Because STSCL logic cannot be used for real ultra-low-voltage operation, the main advantage should be the low power consumption, but the question is if the constant static power consumption of the logic and the power consumption of the bias circuits do not jeopardize this characteristic.

3.1.5.2 Adiabatic Logic

Adiabatic logic for low-power operation has been studied by the group of Schmitt-Landsiedel from the Technical University of Munich. The following information has been summarized from [40]. The idea of *adiabatic* or energy recovering logic is to not use a constant voltage supply, but instead use a pulsed power supply. Moreover, adiabatic logic does not abruptly switch from 0 to V_{dd}, or vice versa, but a voltage ramp is used to charge and recover the energy from the output. A slowly varying voltage source requires less energy to charge a capacitance if its period is longer than the time constant of the charging path. Furthermore, when the supply voltage decreases, the output capacitance is discharged and its stored energy can be recovered by the supply source.

Therefore, adiabatic logic circuits are operated with an oscillating power supply, called the power-clock. Each power-clock cycle consists of four intervals, visualized in Fig. 3.24a. There are four phases of the same power-clock, each shifted 90°. Cascaded logic gates are powered by successive phases ϕ_i of the power-clock. Therefore, adiabatic logic is inherently pipelined. If at a certain location no logic gate is necessary, buffers have to be inserted for synchronization reasons.

Two adiabatic logic families have been found to provide the best energy-efficiency: Positive Feedback Adiabatic Logic (PFAL) and Efficient Charge Recovery Logic (ECRL). Both exhibit a memory functionality, PFAL through a latch element (Fig. 3.24b) and ECRL through a cross-coupled pMOS transistor pair (Fig. 3.24c). According to the authors, the area consumption of adiabatic logic is comparable to standard CMOS logic, but this is only true for complex functions and not for basic functions with a few inputs due to the overhead of the memory functionality.

In the evaluate (E) mode of a logic gate powered by ϕ_0, the outputs are evaluated from the stable input signals (Fig. 3.24a). These outputs are then kept stable for the

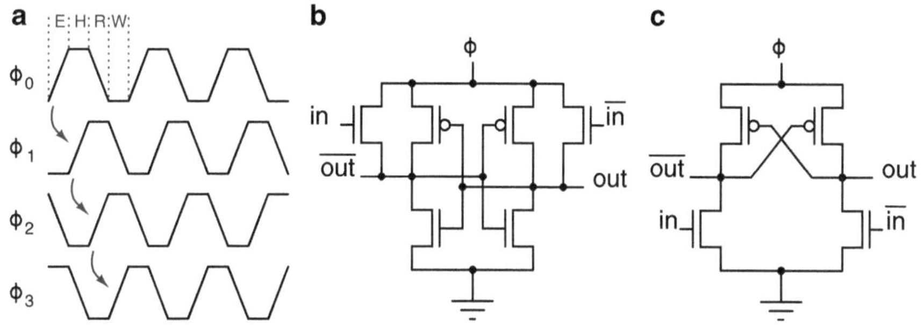

Fig. 3.24 Adiabatic logic: (**a**) four phases of the power-clock, (**b**) PFAL inverter schematic and (**c**) ECRL inverter schematic

subsequent gate in the following mode, i.e. the hold (H) mode. In the recovery (R) mode, energy is recovered by the supply source. The wait (W) mode is inserted for symmetry reasons, as it is easier to generate symmetric signals according to the author.

Adiabatic logic is claimed to save energy compared to standard CMOS logic, but only for moderate operating frequencies. Due to the fact that some energy losses in adiabatic logic are frequency-dependent, there is an optimum frequency for energy-efficiency. For example for a 130 nm CMOS technology, this is supposed to lie around 100 MHz. As for standard CMOS logic, voltage scaling reduces the energy of adiabatic logic. However, the expected energy gain of adiabatic logic compared to standard CMOS logic reduces when lowering V_{dd}. Moreover, there exists a functional supply limit for ECRL and PFAL. The minimum supply is $\max(V_{T,nMOS}, V_{T,pMOS})$ for ECRL and $2 \cdot V_{T,nMOS}$ for PFAL. Below these supply voltages, the circuits malfunction. More information about these lower bounds can be found in [40].

Each adiabatic system consists of two main parts: the digital core design made up of adiabatic gates and the generator of the power-clock signals. An efficient generation of the four phases making up the power-clock is essential to get high energy savings compared to standard CMOS logic with its fixed supply voltage.

Two measured datapath elements have been reported in a 130 nm CMOS technology: an 8-bit ripple carry adder in [5] and a Finite Impulse Response (FIR) filter in [40]. However, both chips have not been measured at frequencies beyond 20 MHz due to test setup limitations, making it difficult to claim that more energy savings would be obtained at 100 MHz. Moreover, both have been measured at quasi-nominal supplies: the adder at 1.2 V and the FIR was reported to function down to 800 mV, which is not spectacularly low. The largest drawback of this adiabatic logic is however that this research group has never measured a full adiabatic system with the power-clock generation on-chip. Since this is essential to evaluate the claimed energy savings, it is unclear if this adiabatic logic really exhibits low-power potential.

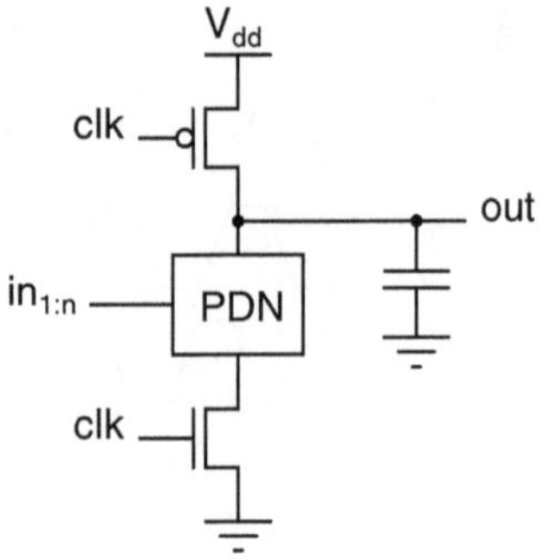

Fig. 3.25 Generic implementation of a dynamic gate

3.1.5.3 Dynamic Logic

As opposed to static logic where the output is always connected to one of the supply rails through a low resistive path, *dynamic* circuits rely on temporary storage of signal values on the capacitance of high-impedance circuit nodes [31]. A dynamic circuit can be obtained by transforming the pMOS load of pseudo-nMOS logic to a clocked pull-up pMOS transistor, as visible in Fig. 3.25. As a result, dynamic operation has two modes, depending on the clock level [46]. When the clock is '0', the output is precharged to '1'. This is called the *precharge* mode. When the clock is '1', the clocked pMOS is turned off and the output may remain high or may be discharged through the PDN, which is the *evaluation* mode. The clocked nMOS *foot* transistor in Fig. 3.25 is optional, depending on whether the input is guaranteed to produce '0' during precharge mode.

Once the output is discharged in the evaluation mode, it cannot be charged again until the next precharge mode. The inputs to the gate can thus make at most one transition during evaluation [31]. Moreover, this must be a low-to-high transition. Therefore, dynamic circuits cannot be cascaded as such, since if their outputs make a transition, it will always be a high-to-low transition. By inserting an inverter after every dynamic gate, this problem can be solved. This is called Domino logic. Consequently, only non-inverting gates can be implemented in Domino logic.

Dynamic logic obtains a similar reduction in transistor count as pseudo-nMOS logic, but avoids the high static power consumption. Furthermore, dynamic logic provides high-speed operation for circuits which are operating at nominal supply voltage. However, it has several disadvantages in ultra-low-voltage operation [45]. Because of the low supply level at which the output will be precharged, only a small amount of charge is stored on the dynamic node. Therefore, this node becomes very sensitive to noise and idle leakage. This is worsened by variations, when for example the precharge pMOS transistor is weakened compared to the PDN. Robustness can therefore not be guaranteed for dynamic circuits in ultra-low-voltage operation.

Sub-threshold dynamic logic, called Sub-Domino logic, has been proposed by Soeleman et al. In [36], simulations in a 0.35 μm CMOS technology showed that Sub-Domino logic was considerably faster and occupied smaller area than standard CMOS logic operating in the sub-threshold region. However, variations have not been studied in this paper, while it is paramount to have a variation-resilient circuit topology for ultra-low-voltage operation. Therefore, it is doubtful that operating dynamic logic at ultra-low supply voltages will provide the required robust functionality.

3.2 Chosen Circuit Topologies

This section discusses the chosen circuit topologies which are used in the ultra-low-voltage prototypes of Chaps. 5 and 6.

3.2.1 *Logic Gates*

The topology used for logic gates is a crucial choice in ultra-low-voltage design to ensure their efficient functionality. It is critical in terms of variation-resilience, energy consumption and speed. From the extensive comparison carried out in Sect. 3.1.4, TG logic has been chosen as preferred topology for logic gates operating in the ultra-low-voltage region. The main reasons for this choice are the inherent robustness of TG logic and its low contribution to leakage. The variation-resilience of TG logic arises from the inclusion of both nMOS and pMOS transistors in each conducting path. Statistically, the effect of variations on both transistors tends to be compensated by the presence of the complementary transistor. The leakage power consumption of TG logic is very low because it does not have direct leakage paths from the supply to the ground.

Another advantage of TG logic is that it uses considerably smaller transistor dimensions compared to standard CMOS logic while achieving better variation-resilience. Moreover, upsizing is often necessary to reduce the sensitivity to variations of ultra-low-voltage standard CMOS logic [22, 29]. These extra margins are not necessary for TG logic. TG design also avoids pMOS stacking and does not require body biasing.

To conclude, TG logic is the most attractive solution for ultra-low-voltage logic gates taking variability into account. Consequently, TG logic is the building block for all logic gates.

Figure 3.26 shows the schematic of the employed TG logic. With this generic logic block, it is possible to construct all 2-input logic gates. Only the order of the inputs needs to be changed to achieve a different logic functionality, as can be seen in Fig. 3.26. For example, a 2-input OR gate requires two inputs A and B and the supply voltage V_{dd}, while its differential equivalent, the NOR gate, has the same inputs at the transistors' gates, but the complementary inputs $\overline{A}$ and V_{ss} at their sources. In this manner, all 2-input logic gates can be constructed (OR, NOR, AND,

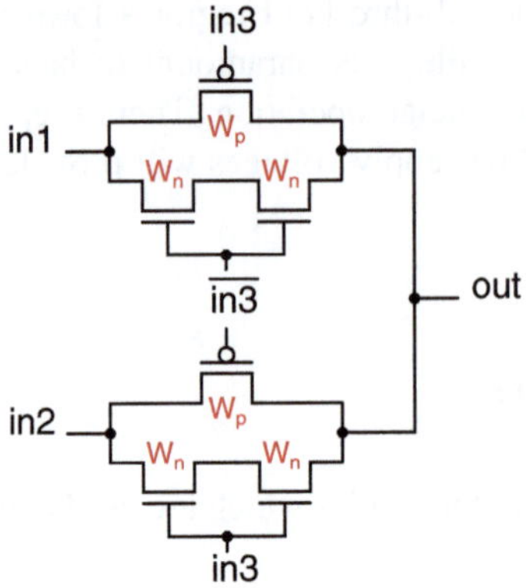

Logic gate	in1	in2	in3
OR	A	V_{dd}	B
NOR	$\overline{A}$	V_{ss}	B
AND	A	V_{ss}	$\overline{B}$
NAND	$\overline{A}$	V_{dd}	$\overline{B}$
XOR	A	$\overline{A}$	B
XNOR	$\overline{A}$	A	B
MUX	A	B	Sel
MUX bar	$\overline{A}$	$\overline{B}$	Sel

Fig. 3.26 Preferred TG logic gate topology: (*left*) schematic of a generic logic gate and (*right*) inputs required to implement the feasible logic functionality

NAND, XOR, XNOR), as well as the 3-input MUX and its differential equivalent. Moreover, with TG logic non-inverted gates like AND and OR gates are possible, which is not the case in standard CMOS logic.

In other words, the design and layout of these logic gates is simplified to the design and layout of just one generic logic block. This modular design considerably simplifies the design of a library of logic gates. The design of this generic block has to be optimized only once for the specific technology at hand, using techniques such as sizing for optimal noise margins and transistor stacking for leakage reduction. The fact that all logic gates have the exact same layout is also beneficial for mismatch.

TG logic requires differential input signals, but the pipelined architecture which is used in the prototypes provides these signals in an efficient way, as will be discussed in Chap. 4.

3.2.2 *Inverter*

Transmission gates are, unfortunately, not ideal switches because they have a series resistance associated with them [31]. Such logic gates cannot be infinitely cascaded since TG logic suffers from some signal loss at the output. By cascading too many logic gates, the robustness can be deteriorated because of too large output signal losses. It is thus necessary to regenerate intermediate signal levels. This regeneration can be performed by inverters or memory elements, such as latches or flip-flops. The inverter topology will be discussed in this section, while memory elements are examined in Sect. 3.3.

Figure 3.27 shows the preferred inverter topology. It consists of a standard CMOS inverter extended with nMOS stacking to relax pMOS sizing, as presented in

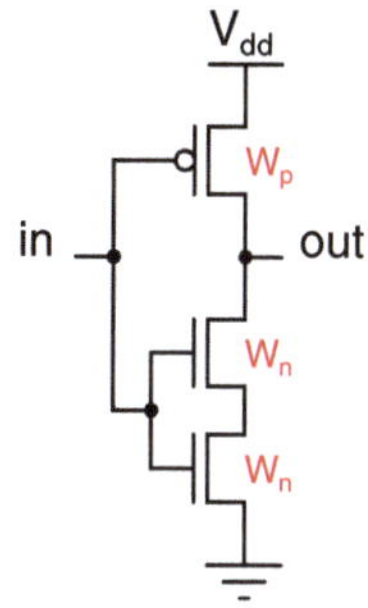

Fig. 3.27 Schematic of the preferred inverter topology

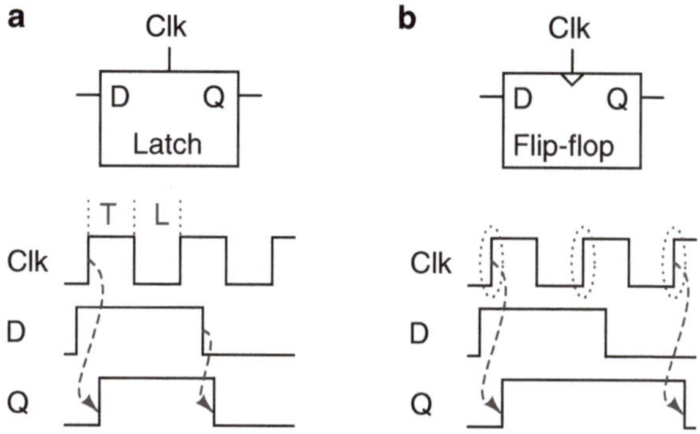

Fig. 3.28 Functionality of (**a**) a level-sensitive latch versus (**b**) an edge-triggered flip-flop

Sect. 3.1.1.4. This type of inverter is preferred to a regular standard CMOS inverter because of the reduced area and the increased variation-resilience.

Remember that the reasons to use nMOS stacking differ from TG logic to this preferred inverter topology. In TG logic, the primary reason is the reduction of the *off*-current $I_{\mathrm{off,nMOS}}$, whereas nMOS stacking is employed in the inverter primarily to decrease the *on*-current $I_{\mathrm{on,nMOS}}$.

3.3 Memory Elements

There are two important types of basic memory elements: a *latch* and a *flip-flop*. Both can be used to store information and are controlled by a clock signal. Figure 3.28 visualizes their functionality. Flip-flops are *edge-triggered*, i.e. when the clock makes a low-to-high transition, the input is copied to the output. The output is stored until the next rising clock edge. On the other hand, latches work in two phases. When the clock is high, the latch is transparent (T) and the data at the input propagates through to the output. When the clock is low, the latch is locked (L) and the output retains the value it last had when transparent. A latch is therefore said to be *level-sensitive*. The implementation of both elements in the ultra-low-voltage region will now be discussed.

Fig. 3.29 Generic schematic of a single-input, differential-output latch

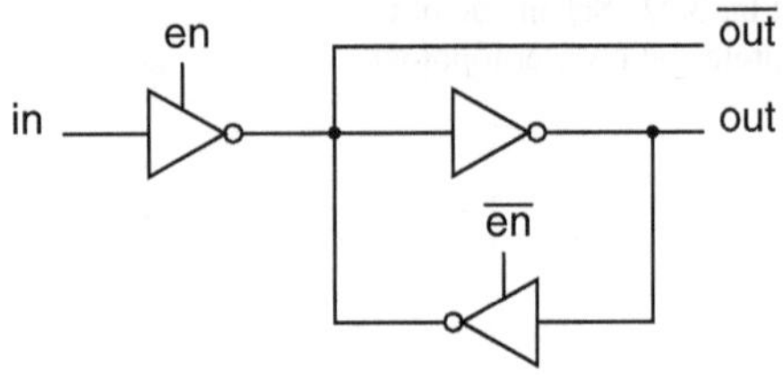

Both memory elements are clocked. As a result, when they are used in an architecture, this architecture becomes pipelined. More information on pipelining will be provided in Chap. 4.

3.3.1 Latch

In order to be able to store data, some form of feedback is necessary. Figure 3.29 shows a generic schematic of a latch with a single input and differential outputs. The cross-coupled inverters provide the feedback functionality. As visible, one of the cross-coupled inverters is a *tristate* inverter, which can be switched on and off. Therefore, the cross-coupling can be turned off. This ratioless behavior is important for robust ultra-low-voltage functionality of the latch. The feedback loop could be implemented in a ratioed fashion as well. However, as already discussed in Sect. 3.1.2, this is undesirable for ultra-low-voltage systems due to their high sensitivity to variations. The interested reader is referred to [8] for a more elaborate discussion on the unsuitability of ratioed latches for sub-threshold operation. Because of the cross-coupled inverters, latches restore the signal levels of their input signals, which is a beneficial characteristic in ultra-low-voltage design.

The clock signals of Fig. 3.28 will from now on be addressed as enable signals, abbreviated to *en*. The latch in Fig. 3.29 consists of a regular inverter and two tristate inverters, controlled by a differential enable signal. The latch functionality of the circuit can easily be verified: in the transparent phase, when *en* is high, the input tristate inverter as well as the regular inverter conduct the input to the output. The feedback path is cut off through the other tristate inverter. In the locked phase, the input tristate inverter is turned off, while the cross-coupled inverters store the data.

Two possible implementations of a tristate inverter are shown in Fig. 3.30: a full-CMOS tristate inverter and a TG-based tristate inverter. In the full-CMOS tristate inverter, the enable transistors are placed in the PUN and PDN of the inverter. However, this introduces pMOS stacking and therefore excessive pMOS sizes are required to ensure good performance in all process corners, as explained in Sect. 3.1.1.3. On the other hand, the TG-based implementation is switched on and off by a transmission gate. When the transmission gate is switched on, both the nMOS and pMOS transistors are turned on. The worst corner for pMOS stacking is the fnsp corner. When the PUN of the inverter is now conducting, the stronger

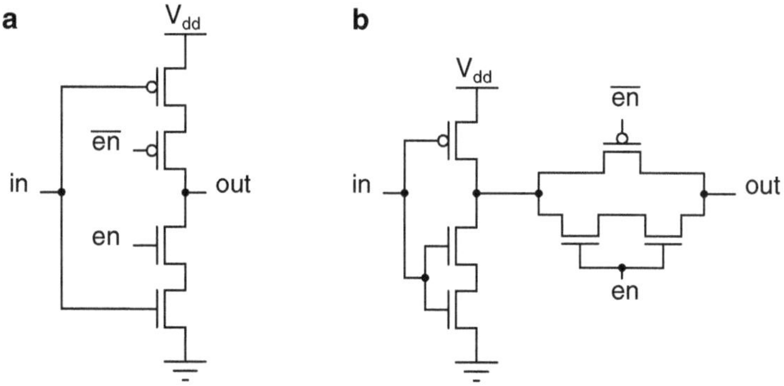

Fig. 3.30 Schematics of (**a**) a full-CMOS tristate inverter and (**b**) a TG-based tristate inverter

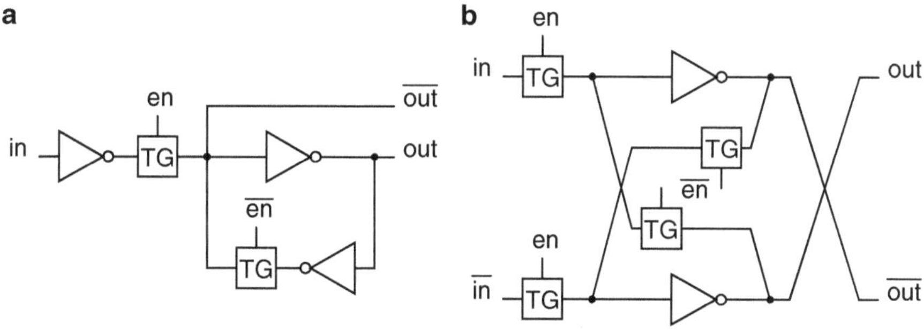

Fig. 3.31 Schematics of differential-output latches with (**a**) a single input and (**b**) differential inputs

nMOS transistors of the transmission gate can compensate the weaker pMOS. Therefore, excessive sizing for process corners is relaxed because the effect of the pMOS stacking is reduced in the TG-based tristate inverter. As a result, the TG-based tristate inverter is preferred for ultra-low-voltage operation because it avoids pure pMOS stacking and hence occupies less area and is more variation-resilient [32].

As explained in Sect. 3.2.1, TG logic requires differential input signals. An attractive property of this latch is that it provides differential output signals in an efficient way, since the complementary output signal is already available without the need for extra circuitry.

In the prototypes which will be presented in Chaps. 5 and 6, two types of latches have been used, as shown in Fig. 3.31. Both have differential outputs which serve as input for the TG logic, but their input signals differ. Figure 3.31a shows a single-input latch [32], while Fig. 3.31b shows a latch with differential inputs [34]. The latter latch can be used when differential input signals are available, i.e. when all TG logic is implemented differentially, whereas the single-input latch can only be used in non-differential cases. In the latches, the same methodology as before has been used: the inverters are implemented as stacked nMOS inverters and the transmission gates which serve as control switches have stacked nMOS transistors as well.

Note that the differential implementation of the latch has a few advantages over the single-input one. First, the number of inverters can be reduced when going from a single input to complementary inputs. This seems counterintuitive, but can be explained by the fact that the inverter at the input is not necessary anymore, since the outputs are still regenerated in both the transparent and the locked phase. On the contrary, if the input inverter of the single-input latch would be removed, $\overline{out}$ would not be amplified through an inverter in the transparent phase. Since the inverters have a significantly higher contribution to leakage than the transmission gate switches, minimizing the number of inverters while ensuring regeneration of the signal levels minimizes leakage. Second, the full differential nature of the latch adds to the variation-resilience of the total design. This is due to the fact that chances are much lower that variations will compromise the correct interpretation of two complementary inputs than of a single input.

3.3.2 Flip-Flop

By cascading two level-sensitive latches, one sensitive on the high level of the clock and the other on the low level of the clock, an edge-triggered flip-flop is constructed. The first latch is then called the master and the second the slave. If flip-flops are used in the prototypes presented in this work, they all exhibit this master-slave configuration. In literature, they are also called *registers*, but throughout this text, the word 'flip-flop' will be used.

3.4 Sizing in Different Prototypes

To summarize, Table 3.3 provides the sizing of the basic building blocks which have been discussed in this chapter for the four different prototypes. These prototypes will be presented in Chaps. 5 and 6.

3.5 Conclusion

This chapter explored the design of gate-level building blocks that can ensure robust operation in the ultra-low-voltage region. These basic building blocks will be used to build the prototypes of Chaps. 5 and 6. The critical factor which was decisive in the evaluation of the circuit topologies has been variation-resilience. As a result,

Table 3.3 Sizing of the basic building blocks in the four prototypes

Prototype		Adder	MAC	MAC	JPEG
CMOS technology		90 nm	90 nm	40 nm	40 nm
Inverter	W_p	6	6	5	5
	W_n	1	1	1	1
Inverter latch	W_p	6	9	10	10
	W_n	1	1.5	2	2
TG logic	W_p	1	2	2	2
	W_n	2	2	1	1
TG latch	W_p	1	1	2	2
	W_n	2	2	1	1

preferred implementations for logic gates, inverters, latches and flip-flops have been achieved. The following chapter will make use of these building blocks when discussing the various architectural sub- and near-threshold trade-offs.

References

1. Akgun O, Rodrigues J, Leblebici Y, Owall V (2012) High-level energy estimation in the sub-Vt domain: simulation and measurement of a cardiac event detector. IEEE Tran Biomed Circuits Syst 6(1):15–27. DOI: 10.1109/TBCAS.2011.2157505
2. Alarcón LP, Liu TT, Pierson MD, Rabaey JM (2007) Exploring very low-energy logic: a case study. J Low Power Electron 3(3):223–233. DOI: 10.1166/jolpe.2007.136
3. Alioto M (2010) Understanding DC behavior of subthreshold CMOS logic through closed-form analysis. IEEE Trans Circuits Syst Regul Pap 57(7):1597–1607. DOI: 10.1109/TCSI.2009.2034233
4. Alioto M (2012) Ultra-low power VLSI circuit design demystified and explained: a tutorial. IEEE Trans Circuits Syst Regul Pap 59(1):3–29. DOI: 10.1109/TCSI.2011.2177004
5. Amirante E, Fischer J, Lang M, Bargagli-Stoffi A, Berthold J, Heer C, Schmitt-Landsiedel D (2003) An ultra low-power adiabatic adder embedded in a standard 0.13 μm CMOS environment. In: Proceedings of the IEEE European solid-state circuits conference (ESSCIRC), pp 599–602. DOI: 10.1109/ESSCIRC.2003.1257206
6. Bol D, Ambroise R, Flandre D, Legat JD (2009) Interests and limitations of technology scaling for subthreshold logic. IEEE Trans Very Large Scale Integr VLSI Syst 17(10):1508–1519. DOI: 10.1109/TVLSI.2008.2005413
7. Bol D, De Vos J, Hocquet C, Botman F, Durvaux F, Boyd S, Flandre D, Legat JD (2013) Sleepwalker: A 25-MHz 0.4-V sub-mm^2 7-μW/MHz microcontroller in 65-nm LP/GP CMOS for low-carbon wireless sensor nodes. IEEE J Solid State Circuits 48(1):20–32. DOI: 10.1109/JSSC.2012.2218067
8. Calhoun B, Chandrakasan A (2006) Ultra-dynamic voltage scaling (UDVS) using sub-threshold operation and local voltage dithering. IEEE J Solid State Circuits 41(1):238–245. DOI: 10.1109/JSSC.2005.859886
9. Chang IJ, Park SP, Roy K (2010) Exploring asynchronous design techniques for process-tolerant and energy-efficient subthreshold operation. IEEE J Solid State Circuits 45(2):401–410. DOI: 10.1109/JSSC.2009.2036764

10. Clerc S, Abouzeid F, Argoud F, Kumar A, Kumar R, Roche P (2011) A 240 mV 1 MHz, 340 mV 10 MHz, 40 nm CMOS, 252 bits frame decoder using ultra-low voltage circuit design platform. In: Proceedings of the IEEE international conference on electronics, circuits and systems (ICECS), pp 117–120. DOI: 10.1109/ICECS.2011.6122228
11. Hanson S, Seok M, Sylvester D, Blaauw D (2008) Nanometer device scaling in subthreshold logic and SRAM. IEEE Trans Electron Devices 55(1):175–185. DOI: 10.1109/TED.2007.911033
12. Hanson S, Zhai B, Seok M, Cline B, Zhou K, Singhal M, Minuth M, Olson J, Nazhandali L, Austin T, Sylvester D, Blaauw D (2008) Exploring variability and performance in a sub-200-mV processor. IEEE J Solid State Circuits 43(4):881–891. DOI: 10.1109/JSSC.2008.917505
13. Henzler S (2007) Power management of digital circuits in deep sub-micron CMOS technologies. Springer, New York
14. Hwang ME, Raychowdhury A, Kim K, Roy K (2007) A 85 mV 40 nW process-tolerant subthreshold 8x8 FIR filter in 130nm technology. In: Proceedings of the IEEE symposium on VLSI circuits (VLSIC), pp 154–155. DOI: 10.1109/VLSIC.2007.4342695
15. Jain S, Khare S, Yada S, Ambili V, Salihundam P, Ramani S, Muthukumar S, Srinivasan M, Kumar A, Gb S, Ramanarayanan R, Erraguntla V, Howard J, Vangal S, Dighe S, Ruhl G, Aseron P, Wilson H, Borkar N, De V, Borkar S (2012) A 280 mV-to-1.2 V wide-operating-range IA-32 processor in 32 nm CMOS. In: Proceedings of the IEEE international solid-state circuits conference (ISSCC), pp 66–68. DOI: 10.1109/ISSCC.2012.6176932
16. Jeon D, Seok M, Chakrabarti C, Blaauw D, Sylvester D (2012) A super-pipelined energy efficient subthreshold 240 MS/s FFT core in 65 nm CMOS. IEEE J Solid State Circuits 47(1):23–34. DOI: 10.1109/JSSC.2011.2169311
17. Jocke SC, Bolus J, Wooters S, Jurik A, Weaver A, Blalock T, Calhoun B (2009) A 2.6-μW subthreshold mixed-signal ECG SoC. In: Proceedings of the IEEE symposium on VLSI circuits (VLSIC), pp 60–61
18. Kao J, Miyazaki M, Chandrakasan A (2002) A 175-mV multiply-accumulate unit using an adaptive supply voltage and body bias architecture. IEEE J Solid State Circuits 37(11):1545–1554. DOI: 10.1109/JSSC.2002.803957
19. Kaul H, Anders M, Mathew S, Hsu S, Agarwal A, Krishnamurthy R, Borkar S (2008) A 320 mV 56 μW 411GOPS/Watt ultra-low voltage motion estimation accelerator in 65nm CMOS. In: Proceedings of the IEEE international solid-state circuits conference (ISSCC), pp 316–317. DOI: 10.1109/ISSCC.2008.4523184
20. Klinefelter A, Zhang Y, Otis B, Calhoun B (2012) A programmable 34 nW/channel subthreshold signal band power extractor on a body sensor node SoC. IEEE Trans Circuits Syst Express Briefs 59(12):937–941. DOI: 10.1109/TCSII.2012.2231041
21. Konijnenburg M, Cho Y, Ashouei M, Gemmeke T, Kim C, Hulzink J, Stuyt J, Jung M, Huisken J, Ryu S, Kim J, de Groot H (2013) Reliable and energy-efficient 1 MHz 0.4 V dynamically reconfigurable SoC for ExG applications in 40 nm LP CMOS. In: Proceedings of the IEEE international solid-state circuits conference (ISSCC), pp 430–431. DOI: 10.1109/ISSCC.2013.6487801
22. Kwong J, Chandrakasan A (2006) Variation-driven device sizing for minimum energy subthreshold circuits. In: Proceedings of the ACM/IEEE international symposium on low power electronics and design (ISLPED), pp 8–13. DOI: 10.1109/LPE.2006.4271799
23. Kwong J, Ramadass Y, Verma N, Chandrakasan A (2009) A 65 nm sub-Vt microcontroller with integrated SRAM and switched capacitor DC-DC converter. IEEE J Solid State Circuits 44(1):115–126. DOI: 10.1109/JSSC.2008.2007160
24. Lutkemeier S, Jungeblut T, Berge H, Aunet S, Porrmann M, Ruckert U (2013) A 65 nm 32b subthreshold processor with 9T multi-Vt SRAM and adaptive supply voltage control. IEEE J Solid State Circuits 48(1):8–19. DOI: 10.1109/JSSC.2012.2220671
25. Makipää J, Turnquist MJ, Laulainen E, Koskinen L (2012) Timing-error detection design considerations in subthreshold: an 8-bit microprocessor in 65 nm CMOS. J Low Power Electron Appl 2(2):180–196. DOI: 10.3390/jlpea2020180

26. Narendra S, Borkar S, De V, Antoniadis D, Chandrakasan A (2001) Scaling of stack effect and its application for leakage reduction. In: Proceedings of the ACM/IEEE international symposium on low power electronics and design (ISLPED), pp 195–200. DOI: 10.1109/LPE.2001.945400
27. Narendra S, Keshavarzi A, Bloechel B, Borkar S, De V (2003) Forward body bias for microprocessors in 130-nm technology generation and beyond. IEEE J Solid-State Circuits 38(5):696–701. DOI: 10.1109/JSSC.2003.810054
28. Narendra S, Chandrakasan A (2006) Leakage in nanometer CMOS technologies. Springer, Berlin
29. Pu Y, Pineda de Gyvez J, Corporaal H, Ha Y (2007) Vt balancing and device sizing towards high yield of sub-threshold static logic gates. In: Proceedings of the ACM/IEEE international symposium on low power electronics and design (ISLPED), pp 355–358. DOI: 10.1145/1283780.1283857
30. Pu Y, Pineda de Gyvez J, Corporaal H, Ha Y (2010) An ultra-low-energy multi-standard JPEG co-processor in 65 nm CMOS with sub/near threshold supply voltage. IEEE J Solid State Circuits 45(3):668–680. DOI: 10.1109/JSSC.2009.2039684
31. Rabaey J, Chandrakasan A, Nikolic B (2003) Digital integrated circuits: a design perspective, 2nd edn. Prentice Hall, Englewood Cliffs
32. Reynders N, Dehaene W (2011) A 190 mV supply, 10 MHz, 90 nm CMOS, pipelined sub-threshold adder using variation-resilient circuit techniques. In: Proceedings of the IEEE Asian solid-state circuits conference (A-SSCC), pp 113–116. DOI: 10.1109/ASSCC.2011.6123617
33. Reynders N, Dehaene W (2012) Variation-resilient building blocks for ultra-low-energy sub-threshold design. IEEE Trans Circuits Syst Express Briefs 59(12):898–902. DOI: 10.1109/TCSII.2012.2231022
34. Reynders N, Dehaene W (2012) Variation-resilient sub-threshold circuit solutions for ultra-low-power digital signal processors with 10 MHz clock frequency. In: Proceedings of the IEEE European solid-state circuits conference (ESSCIRC), pp 474–477. DOI: 10.1109/ESSCIRC.2012.6341358
35. Soeleman H, Roy K (1999) Ultra-low power digital subthreshold logic circuits. In: Proceedings of the ACM/IEEE international symposium on low power electronics and design (ISLPED), pp 94–96
36. Soeleman H, Roy K, Paul B (2001) Sub-domino logic: ultra-low power dynamic sub-threshold digital logic. In: Proceedings of the IEEE international conference on VLSI design, pp 211–214. DOI: 10.1109/ICVD.2001.902662
37. Sze V, Chandrakasan A (2007) A 0.4-V UWB baseband processor. In: Proceedings of the ACM/IEEE international symposium on low power electronics and design (ISLPED), pp 262–267. DOI: 10.1145/1283780.1283837
38. Tajalli A, Brauer E, Leblebici Y, Vittoz E (2008) Subthreshold source-coupled logic circuits for ultra-low-power applications. IEEE J Solid State Circuits 43(7):1699–1710. DOI: 10.1109/JSSC.2008.922709
39. Tajalli A, Leblebici Y (2010) Extreme low-power mixed signal IC design: subthreshold source-coupled circuits. Springer, New York
40. Teichmann P (2012) Adiabatic logic: future trend and system level perspective. Springer, New York
41. Von Arnim K, Borinski E, Seegebrecht P, Fiedler H, Brederlow R, Thewes R, Berthold J, Pacha C (2005) Efficiency of body biasing in 90-nm CMOS for low-power digital circuits. IEEE J Solid State Circuits 40(7):1549–1556. DOI: 10.1109/JSSC.2005.847517
42. Wang A, Chandrakasan A (2004) A 180 mV FFT processor using subthreshold circuit techniques. In: Proceedings of the IEEE international solid-state circuits conference (ISSCC), pp 292–293. DOI: 10.1109/ISSCC.2004.1332709
43. Wang A, Chandrakasan A (2005) A 180-mV subthreshold FFT processor using a minimum energy design methodology. IEEE J Solid State Circuits 40(1):310–319. DOI: 10.1109/JSSC.2004.837945

44. Wang JS, Li HY, Yeh C, Chen TF (2005) Design techniques for single-low-Vdd CMOS systems. IEEE J Solid State Circuits 40(5):1157–1165. DOI: 10.1109/JSSC.2005.845979
45. Wang A, Calhoun B, Chandrakasan A (2006) Sub-threshold design for ultra low-power systems. Springer, New York
46. Weste N, Harris D (2011) CMOS VLSI design: a circuits and systems perspective, 4th edn. Addison-Wesley, Boston
47. Yano K, Yamanaka T, Nishida T, Saito M, Shimohigashi K, Shimizu A (1990) A 3.8-ns CMOS 16x16-b multiplier using complementary pass-transistor logic. IEEE J Solid State Circuits 25(2):388–395. DOI: 10.1109/4.52161
48. Zhai B, Pant S, Nazhandali L, Hanson S, Olson J, Reeves A, Minuth M, Helfand R, Austin T, Sylvester D, Blaauw D (2009) Energy-efficient subthreshold processor design. IEEE Trans Very Large Scale Integr VLSI Syst 17(8):1127–1137. DOI: 10.1109/TVLSI.2008.2007564
49. Zhou J, Jayapal S, Busze B, Huang L, Stuyt J (2011) A 40 nm inverse-narrow-width-effect-aware sub-threshold standard cell library. In: Proceedings of the ACM/EDAC/IEEE design automation conference (DAC), pp 441–446

Chapter 4
Architectural Design

After introducing the preferred gate-level building blocks for ultra-low-voltage operation in Chap. 3, this chapter will explore various architectural options for the prototypes of this book. By examining their benefits and drawbacks, recommendations will be provided for efficient and robust ultra-low-voltage functionality.

Section 4.1 will start this chapter with theoretical considerations on energy consumption, specifically for transistors operating in the weak inversion region and for circuits which are subjected to high variability.

Section 4.2 will explore architectural consequences of using TG logic. A focus is given to the cascading of logic gates and how this can be implemented taking into account the need for complementary input signals, as well as the necessary regeneration of the voltage levels of the output signals. The advantages and disadvantages of increasing the logic depth will be discussed [4]. To determine the optimal logic depth for ultra-low-voltage operation, a test setup is proposed, and its results are presented [3]. Furthermore, attention will be given to differential TG logic and its consequences.

If the aforementioned regeneration is performed by clocked elements, pipelining is introduced, which is the topic of Sect. 4.3. Various pipelining schemes will be explored to assess their suitability for sub- or near-threshold designs. Several parameters have a large influence on pipelined architectures, resulting in design considerations which are often contradictory. To be able to provide recommendations for the field of interest and possible applications of this book, these considerations will therefore be evaluated carefully.

Furthermore, the design methodology which is based on the conclusions of both this and the previous chapter will be presented in Sect. 4.4. The different steps which are used to design and layout the prototypes of the subsequent chapters will be discussed profoundly.

Finally, Sect. 4.5 will discuss the I/O circuits required for the measurement setup of the prototypes.

N. Reynders, W. Dehaene, *Ultra-Low-Voltage Design of Energy-Efficient Digital Circuits*, Analog Circuits and Signal Processing,
DOI 10.1007/978-3-319-16136-5_4

4.1 Theoretical Considerations

This section will explore theoretical considerations for ultra-low-voltage systems with a focus on energy consumption. It will elaborate on how static and dynamic energy relate to each other, as well as on minimizing total energy consumption. The section will start with the nominal case, and will then extend the discussion to systems in the presence of variations.

The definitions of dynamic energy consumption $E_{\mathrm{dyn},1}$ and static energy consumption $E_{\mathrm{stat},1}$ for a single logic gate are the following:

$$E_{\mathrm{dyn},1} = C \cdot V_{\mathrm{dd}}^2 \tag{4.1}$$

$$E_{\mathrm{stat},1} = I_{\mathrm{off}} \cdot V_{\mathrm{dd}} \cdot t_{\mathrm{d}} \tag{4.2}$$

where t_{d} is the delay which is defined by:

$$t_{\mathrm{d}} = \frac{C \cdot V_{\mathrm{dd}}}{I_{\mathrm{on}} - I_{\mathrm{off}}} \tag{4.3}$$

Assuming that $I_{\mathrm{on}} \gg I_{\mathrm{off}}$ provides an approximation for t_{d}:

$$t_{\mathrm{d}} \approx \frac{C \cdot V_{\mathrm{dd}}}{I_{\mathrm{on}}} \tag{4.4}$$

Evidently, this assumption of I_{on} being considerably higher than I_{off} is true when working in the nominal supply region, but what about the weak inversion region? As discussed before, it is imperative to still have a reasonable $I_{\mathrm{on}}/I_{\mathrm{off}}$ ratio in order to guarantee correct functionality for circuits which are operating at ultra-low supply voltages. Otherwise, circuits will not exhibit correct behavior anymore when subjected to the high variations which exist in the ultra-low-voltage region. Section 2.3.1 proposed a minimum value of 50 for this current ratio. Consequently, the assumption still holds since I_{on} is 50 times higher than I_{off}, and Eq. (4.4) can be used.

Inserting (4.4) in (4.2) gives:

$$E_{\mathrm{stat},1} = C \cdot V_{\mathrm{dd}}^2 \cdot \frac{I_{\mathrm{off}}}{I_{\mathrm{on}}} \tag{4.5}$$

When these energy equations are extended to a digital system with N logic gates and activity α, this gives:

$$E_{\mathrm{dyn}} = \alpha \cdot N \cdot C \cdot V_{\mathrm{dd}}^2 \tag{4.6}$$

$$E_{\mathrm{stat}} = N \cdot C \cdot V_{\mathrm{dd}}^2 \cdot \frac{I_{\mathrm{off}}}{I_{\mathrm{on}}} \tag{4.7}$$

4.1.1 Energy Ratio

As discussed in Sect. 2.2.2, leakage plays a much bigger role in circuit functionality in the weak inversion region than it does in the strong inversion region. Using the derived equations, this can be theoretically studied by examining the ratio of E_{dyn} to E_{stat}:

$$\begin{aligned}\frac{E_{\text{dyn}}}{E_{\text{stat}}} &= \frac{\alpha \cdot N \cdot C \cdot V_{\text{dd}}^2}{N \cdot C \cdot V_{\text{dd}}^2 \cdot \frac{I_{\text{off}}}{I_{\text{on}}}} \\ &= \alpha \cdot \frac{I_{\text{on}}}{I_{\text{off}}}\end{aligned} \tag{4.8}$$

This results in a surprisingly clean relation, with only a few parameters. However, there is a way to define the $E_{\text{dyn}}/E_{\text{stat}}$ ratio so that the current ratio is replaced by technological parameters. Recall the derived Eq. (2.16):

$$\frac{I_{\text{on}}}{I_{\text{off}}} = \exp\left(\frac{V_{\text{dd}}}{n \cdot V_{\text{th}}}\right)$$

By combining this with the definition of the sub-threshold slope S_{S} of (2.19):

$$S_{\text{S}} = n \cdot V_{\text{th}} \cdot \ln(10)$$

This results in:

$$\begin{aligned}\frac{I_{\text{on}}}{I_{\text{off}}} &= \exp\left(\frac{V_{\text{dd}} \cdot \ln(10)}{S_{\text{S}}}\right) \\ &= 10^{\frac{V_{\text{dd}}}{S_{\text{S}}}}\end{aligned} \tag{4.9}$$

Using the resulting Eq. (4.9) leads to the following elegant relation:

$$\frac{E_{\text{dyn}}}{E_{\text{stat}}} = \alpha \cdot 10^{\frac{V_{\text{dd}}}{S_{\text{S}}}} \tag{4.10}$$

Hence, the manner how the dynamic energy relates to the static energy is defined by the activity of the system, the supply voltage at which it is operating and the sub-threshold slope.

Since S_{S} is a technological parameter, it is fixed for a certain CMOS technology. The lower the value of S_{S}, the steeper the slope of I_{ds}, and the higher the $I_{\text{on}}/I_{\text{off}}$ ratio for a certain supply. A lower S_{S} is thus more attractive. Recall that the theoretical lower bound of S_{S} is 60 mV/decade at room temperature. The fact that a reduced S_{S} is more beneficial can be derived from (4.10) as well: the static energy reduces compared to the dynamic energy at lower S_S. However, when designing in a certain technology, the S_S is not a parameter which can be adjusted. What about the other two parameters?

In this work, a certain system with a predefined architecture is assumed. Hence, the activity is a given parameter. As a result, the supply voltage is key to defining how the dynamic energy relates to the static energy consumption. This theoretical derivation reveals that when lowering the supply, the static energy will relatively become more important. In fact, the supply voltage completely defines the E_{dyn}/E_{stat} ratio for a certain system, which will also be shown in the measurement results of the prototypes of Chaps. 5 and 6.

4.1.2 Total Energy Consumption

The total energy consumption E_{tot} is calculated by adding the dynamic and the static components. Taking into account (4.6) and (4.7), E_{tot} of a system becomes:

$$\begin{aligned} E_{tot} &= E_{dyn} + E_{stat} \\ &= N \cdot C \cdot V_{dd}^2 \cdot \left(\alpha + \frac{I_{off}}{I_{on}}\right) \\ &= \alpha \cdot N \cdot C \cdot V_{dd}^2 \cdot \left(1 + \frac{1}{\alpha} \cdot \frac{I_{off}}{I_{on}}\right) \end{aligned} \tag{4.11}$$

By using (4.9), (4.11) can be rewritten:

$$E_{tot} = \alpha \cdot N \cdot C \cdot V_{dd}^2 \cdot \left(1 + \frac{1}{\alpha} \cdot 10^{-\frac{V_{dd}}{S_S}}\right) \tag{4.12}$$

Furthermore, (4.11) can be rewritten in another manner as well, by using the currents for a total system instead of the currents of a single logic gate. These currents $I_{on,tot}$ and $I_{off,tot}$ are defined in a similar manner as the energy consumptions previously:

$$I_{on,tot} = \alpha \cdot N \cdot I_{on} \quad \Leftrightarrow \quad I_{on} = \frac{I_{on,tot}}{\alpha \cdot N} \tag{4.13}$$

$$I_{off,tot} = N \cdot I_{off} \quad \Leftrightarrow \quad I_{off} = \frac{I_{off,tot}}{N} \tag{4.14}$$

Given a certain system, the currents $I_{on,tot}$ and $I_{off,tot}$ can be measured and therefore provide the possibility of validating these theoretical considerations with empirical data later on. Employing these currents results in the following equation:

$$E_{\text{tot}} = \alpha \cdot N \cdot C \cdot V_{\text{dd}}^2 \cdot \left(1 + \frac{1}{\alpha} \cdot \frac{\frac{I_{\text{off,tot}}}{N}}{\frac{I_{\text{on,tot}}}{\alpha \cdot N}}\right)$$

$$= \alpha \cdot N \cdot C \cdot V_{\text{dd}}^2 \cdot \left(1 + \frac{I_{\text{off,tot}}}{I_{\text{on,tot}}}\right) \tag{4.15}$$

When combining (4.12) and (4.15), an equation to calculate the activity α can be derived:

$$\frac{1}{\alpha} \cdot 10^{-\frac{V_{\text{dd}}}{S_S}} = \frac{I_{\text{off,tot}}}{I_{\text{on,tot}}}$$

$$\Updownarrow$$

$$\alpha = \frac{I_{\text{on,tot}}}{I_{\text{off,tot}}} \cdot 10^{-\frac{V_{\text{dd}}}{S_S}} \tag{4.16}$$

This derivation could lead one to believe that by using measurement results for a certain system designed in a specific technology, the activity, generally a parameter which is hard to quantify exactly, could be calculated using (4.16). Unfortunately, this is not the case, since these measurements are performed in an actual system which is subjected to variations, not in a system which exhibits purely nominal behavior. Therefore, it is imperative to take variations into account when coupling these theoretical considerations with measured data.

The question then is where the variability has been neglected so far, by assuming nominal conditions. The delay definition which has been used up till now comprised the delay of a single nominal gate. However, when determining the static energy consumption of a system, this should be the worst-case delay of the slowest gate, as this is the delay which defines the maximal clock frequency of the system:

$$t_{\text{d,worst-case}} = \max_i \left(\frac{C_i}{I_{\text{on},i}}\right) \cdot V_{\text{dd}} \tag{4.17}$$

$$E_{\text{stat}} = N \cdot \max_i \left(\frac{C_i}{I_{\text{on},i}}\right) \cdot V_{\text{dd}}^2 \cdot I_{\text{off}} \tag{4.18}$$

The total energy consumption of (4.11) then becomes:

$$E_{\text{tot}} = \alpha \cdot N \cdot C \cdot V_{\text{dd}}^2 + N \cdot \max_i \left(\frac{C_i}{I_{\text{on},i}}\right) \cdot V_{\text{dd}}^2 \cdot I_{\text{off}} \tag{4.19}$$

This can be reworked to:

$$E_{\text{tot}} = \alpha \cdot N \cdot V_{\text{dd}}^2 \cdot \left(C + \frac{1}{\alpha} \cdot \max_i \left(\frac{C_i}{I_{\text{on},i}}\right) \cdot I_{\text{off}} \cdot \frac{\frac{C}{I_{\text{on}}}}{\frac{C}{I_{\text{on}}}}\right)$$

$$= \alpha \cdot N \cdot C \cdot V_{\mathrm{dd}}^2 \cdot \left(1 + \frac{1}{\alpha} \cdot \frac{\max_i\left(\frac{C_i}{I_{\mathrm{on},i}}\right)}{\frac{C}{I_{\mathrm{on}}}} \cdot \frac{I_{\mathrm{off}}}{I_{\mathrm{on}}}\right) \tag{4.20}$$

The ratio of the maximal over the nominal $\frac{C}{I_{\mathrm{on}}}$ will from now on be addressed as the *variation factor* F_{v}:

$$F_{\mathrm{v}} = \frac{\max_i\left(\frac{C_i}{I_{\mathrm{on},i}}\right)}{\frac{C}{I_{\mathrm{on}}}} \tag{4.21}$$

This variation factor is expected to increase with reducing V_{dd} due to the higher sensitivity of circuit parameters to variations at lower supply voltages. This will be shown later as well but can already be seen by expanding I_{on} with its definition (2.14):

$$F_{\mathrm{v}} = \frac{\max_i\left(\frac{C_i}{I_0 \cdot \exp\left(\frac{V_{\mathrm{dd}} - V_{\mathrm{T},i}}{n \cdot V_{\mathrm{th}}}\right)}\right)}{\frac{C}{I_0 \cdot \exp\left(\frac{V_{\mathrm{dd}} - V_{\mathrm{T}}}{n \cdot V_{\mathrm{th}}}\right)}} \tag{4.22}$$

The absolute V_{T} variation stays the same, but becomes relatively much more important at a supply value near the value of V_{T}. The factor F_{v} thus provides a measure of the amount of variations that is affecting the timing of a system. Equivalently, $1/F_{\mathrm{v}}$ gives a measure of how well the delays of the N logic gates of the system are balanced.

Continuing the calculations on (4.20) by again introducing $I_{\mathrm{on,tot}}$ and $I_{\mathrm{off,tot}}$ from (4.13) and (4.14) leads to:

$$\begin{aligned} E_{\mathrm{tot}} &= \alpha \cdot N \cdot C \cdot V_{\mathrm{dd}}^2 \cdot \left(1 + \frac{1}{\alpha} \cdot F_{\mathrm{v}} \cdot \frac{\frac{I_{\mathrm{off,tot}}}{N}}{\frac{I_{\mathrm{on,tot}}}{\alpha \cdot N}}\right) \\ &= \alpha \cdot N \cdot C \cdot V_{\mathrm{dd}}^2 \cdot \left(1 + F_{\mathrm{v}} \cdot \frac{I_{\mathrm{off,tot}}}{I_{\mathrm{on,tot}}}\right) \end{aligned} \tag{4.23}$$

Taking into account variations, (4.23) can now be combined with (4.12) without any issues:

$$\begin{aligned} \frac{1}{\alpha} \cdot 10^{-\frac{V_{\mathrm{dd}}}{S_{\mathrm{S}}}} &= F_{\mathrm{v}} \cdot \frac{I_{\mathrm{off,tot}}}{I_{\mathrm{on,tot}}} \\ &\Updownarrow \\ F_{\mathrm{v}} \cdot \alpha &= \frac{I_{\mathrm{on,tot}}}{I_{\mathrm{off,tot}}} \cdot 10^{-\frac{V_{\mathrm{dd}}}{S_{\mathrm{S}}}} \end{aligned} \tag{4.24}$$

The right side of this equation consists of known parameters which are fixed by technology or which can be measured. However, the two parameters on the left side of the equation are unknown. Determining the activity factor α of a system is difficult, and the same holds for F_v. What can then be learned from this equation? Since α is constant when sweeping the supply voltage, it could be omitted when normalizing F_v, since $F_{v,norm}$ is displayed in an arbitrary unit in that case.

This leaves the equation to derive $F_{v,norm}$ with only known parameters. The absolute value of $F_{v,norm}$ is then not useful anymore, but relatively it can provide insight into the amount of variations. As a case study, the second and third prototype of this book, which will be presented in Chap. 5, are used. These two prototypes consist of the same Multiply-Accumulate Unit (MAC), but are processed in different technologies, i.e. the 90 nm and the 40 nm CMOS technologies at hand. Since it consists of the same system with identical architectures, their measurement results can be compared in order to study the impact of technology scaling. Currently, the interesting property of those measurements is that the activity of both systems is equal. Therefore, the measured currents as function of V_{dd} can be used to interpret $F_{v,norm}$.

Figure 4.1 shows the resulting calculated $F_{v,norm}$ for both technologies. Note that the normalization has been carried out at a supply voltage of 290 mV, which is the highest supply voltage at which the two designs will be compared in Chap. 5. This results in a somewhat skewed image, since the absolute value of F_v is not expected to be equal at that point. However, the normalization has to be carried out at some voltage, and the highest voltage is the most adequate, since the region of interest of this book is located at the lower supply voltages. Moreover, it is the slope of $F_{v,norm}$ which is interesting to examine.

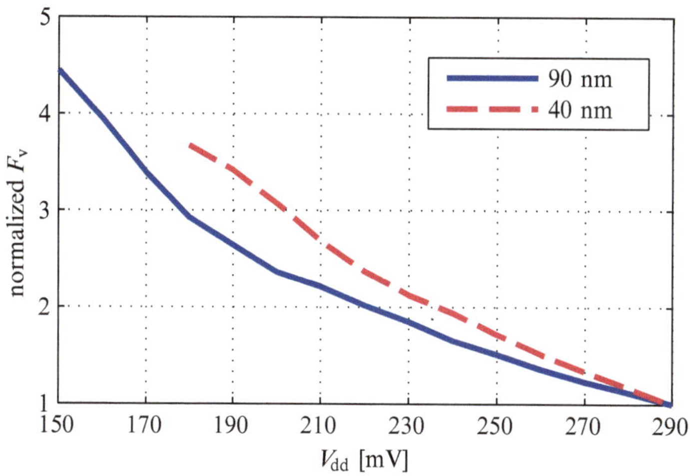

Fig. 4.1 Normalized F_v for the measured MAC prototypes in the technologies at hand

As visible, when reducing the supply voltage, $F_{v,norm}$ increases more for the 40 nm MAC than for the 90 nm MAC. This indicates that the 40 nm version suffers more from variability. Or, equivalently, that the logic gates of the 40 nm MAC are less balanced for lower supplies than the ones of the 90 nm system.

Section 2.3.2 already explained that variability will increase with CMOS technology scaling. However, it remains difficult to observe the increased variability in more advanced technologies during measurements. As will be seen later, the minimum functional supply voltage of the 90 nm MAC is lower than the one of the 40 nm MAC. This suggests that the variability is higher in the 40 nm case, as they were designed with the same design methodology, but this is quite intuitive. Figure 4.1 provides a better, mathematical method to observe this increased variability. The theoretical derivation of this section is validated with measurement results, which produce realistic results. These results clearly provide insight in the phenomenon of increased variability. The remainder of this chapter will explore the architectural design of digital systems from a circuit perspective.

4.2 Cascading Logic Gates

In Chap. 3, the preferred topology for logic gates operating at ultra-low supply voltages has been discussed. In this section, the trade-offs that accompany the use of Transmission Gate (TG) logic on architectural level will be discussed.

4.2.1 Concept

As explained in Sect. 3.2, TG logic suffers from some signal loss at the output and therefore some form of regeneration of intermediate signal levels is necessary. A possibility is to put a regeneration element after each TG logic gate, but of course these elements contribute to the overall leakage power, dynamic energy consumption and delay of the total system. It is more beneficial to cascade multiple logic gates and only regenerate after a certain number of logic gates. The amount of logic gates between two regeneration elements is called the *logic depth*. Cascading logic gates has some benefits, but also some drawbacks. Both will now be discussed.

Several options are possible to implement the regeneration elements, e.g. inverters, latches and flip-flops. In this section, no specification of the used implementation is necessary and therefore the explanation will discuss these elements from a generic perspective.

An important consequence of using TG logic is that cascading multiple gates has a significant impact on the system's architecture. As discussed in Sect. 3.2.1, TG logic requires differential inputs. If a regeneration element is added behind every TG logic gate, this element can provide the differential inputs for the subsequent

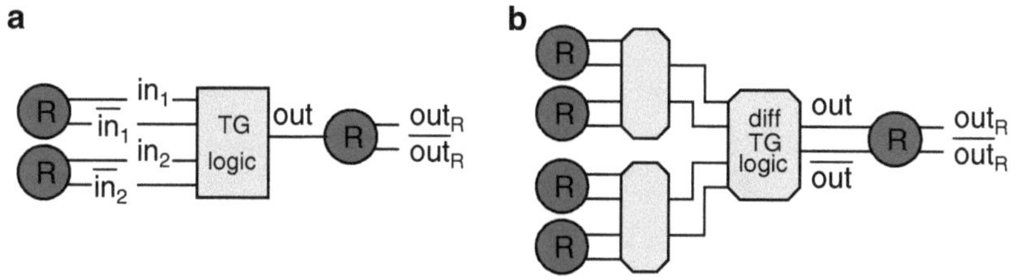

Fig. 4.2 Basic structure of (**a**) single-ended TG logic and (**b**) cascaded differential TG logic, with regeneration elements

logic gate, as visible in Fig. 4.2a. However, if there are multiple cascaded logic gates before a regeneration element, the logic gates should be implemented differentially. An example of two cascaded differential TG logic gates is shown in Fig. 4.2b. The regeneration elements then still have to provide differential outputs, but also receive differential inputs, which can be beneficial. For example, Sect. 3.3.1 discussed how the fully differential implementation of the latch results in reduced leakage and increased variation-resilience. More information on differential TG logic will be provided in Sect. 4.2.3.

The main advantage of cascading logic gates is that the timing variations of these gates are averaged. Assuming Gaussian distributions, the mean of the overall timing variation μ_{cascaded} increases with the number of consecutive gates n, whereas the standard deviation σ_{cascaded} only increases with the square root of the logic depth:

$$\mu_{\text{cascaded}} = n \cdot \mu_{\text{single}} \tag{4.25}$$

$$\sigma_{\text{cascaded}} = \sqrt{n} \cdot \sigma_{\text{single}} \tag{4.26}$$

Therefore, by increasing the logic depth, averaging of timing variations is obtained. As an example, Fig. 4.3 shows the CDF of the propagation delay of different logic depths, i.e. no cascading (depth 1) and two cascaded gates (depth 2), obtained with extensive MC simulations. The standard deviation σ_1 is 3.29 ns, whereas σ_2 is 5.28 ns. The standard deviation thus increases with a factor of 1.6, slightly higher than the square root of 2, which is 1.41, but still significantly lower than 2. As can be seen, the normal distributions fit the data sets reasonably well. However, since the logic gates are operating in the weak inversion region, it is probable that the propagation delay is not really normally distributed, which explains the factor deviation from the square root. Nonetheless, increasing the logic depth in sub-threshold designs proves to be valuable because an important degree of averaging is still obtained [4].

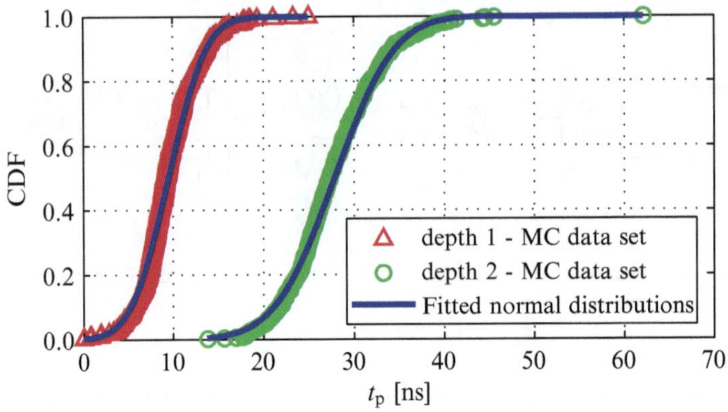

Fig. 4.3 CDF of the propagation delay of cascaded TG logic gates with different logic depths at $V_{dd} = 150\,\text{mV}$, obtained with extensive MC simulations and fitted with normal distributions

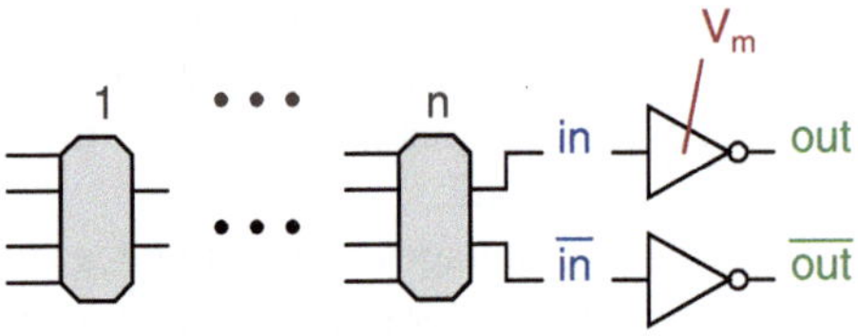

Fig. 4.4 Test setup of a cascade of *n* logic gates, followed by inverters

4.2.2 *Trade-Off*

While the main benefit of cascading logic gates is the introduced averaging of timing variations, some other benefits also exist. The TG logic consumes considerably smaller leakage power and dynamic energy than the regeneration elements. Therefore, maximizing the logic depth while guaranteeing functionality is beneficial in terms of energy consumption. In terms of area, TG logic is sized almost minimally, whereas the regeneration elements require a much larger area.

However, Figs. 3.20 and 3.21 showed that signal losses are present in subthreshold logic gates. As a result, by cascading too many logic gates, the robustness can be deteriorated because of too large output signal losses. Moreover, the regeneration element before the cascade has to be able to drive all TGs. Increasing the logic depth is not infinitely possible with TG logic operating at ultra-low supply voltages. Taking intra-die variations into account, the logic depth is determined by whether or not the regeneration element at the end of the cascade will be able to regenerate the output signals correctly or if it will fail to do so.

Figure 4.4 shows the test setup used to quantify how many logic gates can be cascaded without compromising functionality [3]. The inverter is used as regeneration element, since both the latch and the flip-flop also make use of this inverter. Figure 4.5 then provides the simulation results for logic depths from 1 to

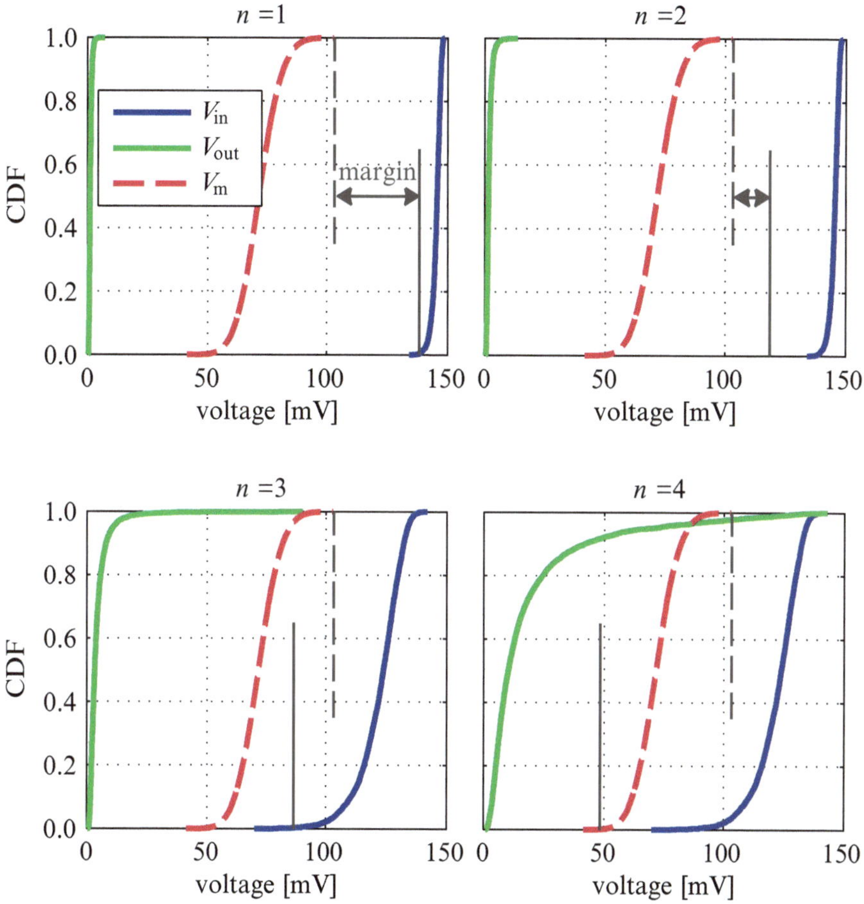

Fig. 4.5 Simulation results from the test setup in Fig. 4.4: CDF of the output voltage level of a cascade of n logic gates V_{in}, compared to the switching point of the inverter V_M and the inverter's output V_{out}, obtained with extensive MC simulations for $V_{dd} = 150\,\text{mV}$

4 under intra-die variations. Since TG logic suffers significantly more from output losses on logic high level (as shown in Fig. 3.21), the figures only display the most pessimistic case where the output of the TG logic gates is high.

Signal losses do not necessarily pose a threat for functionality, as long as the inverter is able to interpret the logic level correctly. Therefore, the spread of the input voltage level V_{in} of the inverter (i.e. the output voltage level of the cascaded logic) is compared to the spread of the switching point V_M of the inverter (see Fig. 4.4). In the case where the input should be a logic '1' but the input voltage of the inverter is lower than its switching voltage, the input is wrongly propagated. To evaluate the chance of this worst-case scenario, a criterion with a yield of 1 wrong propagation out of a billion propagations has been used. If for n logic gates, the chance on wrong propagation is smaller than 1 out of a billion, cascading n gates is considered to not compromise robustness. This yield $Y_{1/1\text{billion}}$ is equal to $(1 - normcdf(6))$ for a single Gaussian distribution, thereby corresponding to 6σ. But, for two uncorrelated distributions, the yield $Y_{1/1\text{billion}}$ is equal to $(1 - normcdf(4))^2$, hence corresponding to 4σ for each distribution. Therefore, both distributions of the output level of the cascade V_{in} and the switching point V_{M} of the inverter are evaluated at 4σ: the vertical lines on Fig. 4.5 indicate the values of $\mu - 4\sigma$ of V_{in} and $\mu + 4\sigma$ of V_{M}. If the two values coincide, a yield of 1 wrong propagation out of a billion propagations is reached. If the first one is higher than the second, there is still margin, as indicated on the figure.

To make sure the cascade of logic gates functions under all circumstances, the simulation is carried out for the target minimum supply voltage of the building blocks, i.e. 150 mV, which was the target $V_{\text{dd,min}}$ of the second prototype. Figure 4.5 demonstrates that for logic depths of 1 and 2 there is still margin on the $Y_{1/1\text{billion}}$-criterion because $\mu_{V_{\text{in}}} - 4\sigma_{V_{\text{in}}} > \mu_{V_{\text{M}}} + 4\sigma_{V_{\text{M}}}$. Consequently, cascades of 1 or 2 logic gates have a chance on wrong propagation that is smaller than 1 out of a billion. The results also show that for cascades of 3 and 4 logic gates $\mu_{V_{\text{in}}} - 4\sigma_{V_{\text{in}}} < \mu_{V_{\text{M}}} + 4\sigma_{V_{\text{M}}}$. Therefore, cascading more than 2 logic gates results in a deteriorated yield.

It can also be seen that the output of an inverter after a cascade of 4 logic gates displays a higher spread than the spread of its input, thereby making it highly discouraging to use such logic depth. To conclude, for a 150 mV supply, the criterion indicates that for a cascade of 2 logic gates the trade-off between energy consumption and guaranteed robustness is optimal. However, for a higher supply voltage, the maximum logic depth to guarantee reliable operation also increases. A reduction of the number of regeneration elements will thus be obtained by redoing the analysis with a higher target supply voltage (e.g. 200 mV) to increase the maximum logic depth. To summarize, depending on the technology at hand and the target minimum supply of a design, performing this analysis results in clear design decisions for the allowable amount of cascading.

Hence, there exists a trade-off to determine the optimal logic depth. One the one hand, by cascading many logic gates, timing variations are averaged, which is very beneficial because of the high timing variations when operating at ultra-low supply voltages. On the other hand, it is imperative that robustness remains guaranteed and that the overall signal loss is restricted by limiting the number of cascaded logic gates.

4.2.3 *Differential TG Logic*

As explained earlier, implementing TG logic differentially allows cascading multiple logic gates. However, differential TG logic has other consequences as well. Firstly, it improves gate reliability because two complementary outputs are available

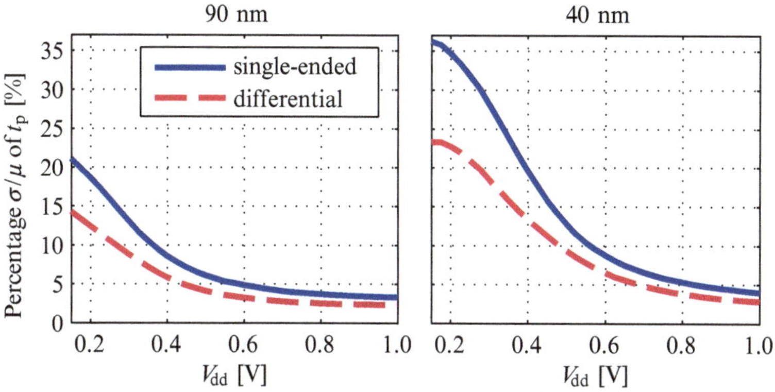

Fig. 4.6 Variation of t_p for a single-ended versus a differential XOR gate as function of V_{dd}, obtained with MC simulations. Results are shown for both technologies at hand: the 90 nm (*left*) and the 40 nm (*right*) CMOS technologies

for interpretation by the regeneration elements, instead of a single output. Secondly, it adds significantly to the variation-resilience of the total design. Figure 4.6 visualizes this effect. A single-ended XOR gate is compared to a differential implementation of a XOR gate when applying intra-die variations. The percentage variation of the propagation delay is considerably lower for the differential XOR than for the single-ended version. As can be seen, this is especially true for very low supply voltages.

Figure 4.6 also shows the difference in variation between both technologies at hand. The increased variations when going to advanced nanometer CMOS technologies (covered in Sect. 2.3.2) are clearly visible. Naturally, the differential version is more variation-resilient in both cases.

Compared to standard CMOS logic, TG logic is much more suited to be implemented differentially, since TG logic already requires complementary input signals. Moreover, using differential logic has the advantage that complex gates such as XOR and XNOR gates can be realized efficiently with a small number of transistors. Of course, wiring complexity does increase with differential logic.

Constructing logic gates differentially does not necessarily increase total area or energy consumption. This seems counterintuitive and is of course not generally true, but in the case of the ultra-low-voltage prototypes of Chaps. 5 and 6 it is related to the architectural and topology choices, as discussed in the previous chapter and in this chapter. The differential implementation of the logic gates allows increasing the logic depth and therefore decreasing the number of regeneration elements. Since the area, leakage power and dynamic energy of the regeneration elements are significantly higher than of a TG logic gate, using differential TG logic with higher logic depth actually improves area density and energy figures.

Table 4.1 Cascading options used in the four prototypes

Prototype	Adder	MAC	MAC	JPEG
CMOS technology	90 nm	90 nm	40 nm	40 nm
Differential TG logic?	✗	✓	✓	✓
Logic depth	1	2	2	3

4.2.4 Realization

The cascading options used in the different prototypes of this book are shown in Table 4.1. The adder in the 90 nm CMOS technology has been implemented with single-ended TG logic, and therefore only has a logic depth of 1.

To increase the logic depth of the MAC in the same 90 nm technology, differential TG logic was used. Extensive MC simulations concluded that the optimal balance for the cascading trade-off was a maximum number of 2 cascaded TG logic gates, for a target minimum supply of 150 mV. Since the MAC in the 40 nm CMOS technology was specifically designed to study the impact of CMOS technology scaling, the same logic depth was used. Performing the simulation analysis for a logic depth of 2 in the 40 nm technology showed that such a logic depth was feasible for a supply voltage around 200 mV, which was satisfactory for the design.

The JPEG encoder in the 40 nm technology on the other hand was designed to operate at a higher minimum supply. Therefore, the analysis to check whether the output of a cascade of TG logic gates can still be interpreted correctly by the following regeneration element resulted in an optimal balance for the trade-off with a cascade of maximally 3 TG logic gates.

4.3 Pipelining

The required regeneration for TG logic can be performed by inverters, latches and flip-flops. The latter two are clocked elements and were discussed in Sect. 3.3. Thereby, when they are used in an architecture, this architecture becomes pipelined. This section will discuss pipelining in general, as well as its consequences and the design considerations which are important for ultra-low-voltage design.

4.3.1 Concept

A pipelined architecture is an architecture where the combinational logic is subdivided into pipeline stages by inserting clocked memory elements in between those stages, as can be seen in Fig. 4.7. Some important pipelining concepts are:

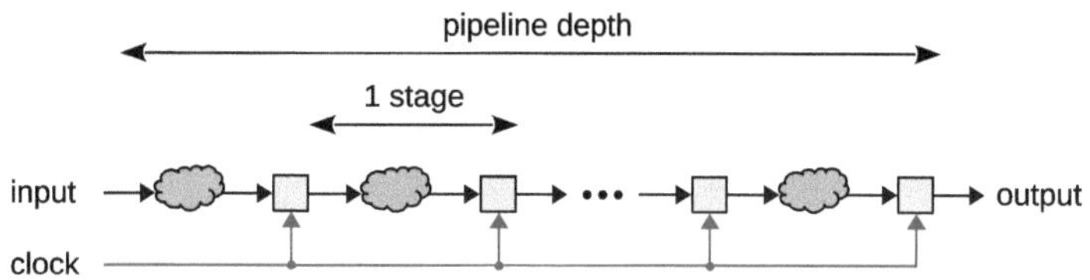

Fig. 4.7 Overview of a pipelined architecture

- *Pipeline stage*: the system's total combinational logic is divided into pipeline stages. One pipeline stage consists of (cascaded) logic followed by a clocked memory element.
- *Pipeline stage length*: the pipeline stage length is the length of a single pipeline stage, and is determined by the logic depth of the combinational logic in that stage.
- *Pipeline depth*: the pipeline depth is the number of pipeline stages in the total pipeline. Given a specified datapath structure, the pipeline depth can be decreased by increasing the pipeline stage length, and vice versa.
- *Latency*: the latency is the amount of time to perform a single computation.
- *Throughput*: throughput is the rate at which data can enter the pipeline, or in other words, the number of computations that are completed in a certain span of time, e.g. in a second.

Pipelining improves resource utilization because it allows more logic gates to perform a useful computation at the same moment. It is a technique which improves the overall processing performance of a certain system due to the simultaneous execution of multiple computations.

By employing pipelining, the clock speed of the system, i.e. the throughput, can be increased. However, the execution time of a stage increases slightly due to the pipelining overhead, and thus the latency increases. To summarize, pipelining trades latency for throughput. Both latency and throughput increase when introducing pipelining. Therefore, pipelining works very effectively for throughput-constrained designs, where a certain data rate is required, but not for latency-constrained circuits, where a certain computation has to finish within a given time frame. Many throughput-constrained scenarios exist, for example the signal processing applications which are the focus of this book.

4.3.2 Benefits and Drawbacks

In the ultra-low-voltage context, pipelining can increase the throughput significantly and therefore it can ensure a higher sub- or near-threshold speed performance. Due to their inherently low logic speed, pipelining is very valuable for ultra-low-voltage designs.

By using pipelining, the clock period can be reduced as long as it is still equal or larger than the delay of the slowest pipeline stage. Therefore, it is recommended to carefully balance all the pipeline stages. This is even more important in ultra-low-voltage designs because of the very high timing variations. The more equalized the nominal stage delays are, the less impact these timing variations will have. Of course, in the end, the period of the clock signal will still be determined by the worst-case stage delay.

Naturally, introducing pipelining in a system adds to the complexity of a design. However, when using TG logic, regeneration is always required after a certain maximum logic depth. Therefore, not much extra effort is necessary to introduce pipelining, while it exhibits some attractive benefits.

4.3.3 Pipelining Schemes

Pipelined architectures can be based on latches or flip-flops, resulting in two different pipelining schemes. Recall that flip-flops are edge-triggered and latches are level-sensitive. In a system using flip-flops, the input data must be ready on the rising (or the falling) edge of the clock. If the data arrives late, the circuit produces the wrong result. If the data arrives early, the time between arrival and propagation goes unused [6]. In contrast, when a system uses latches, the data can arrive in the entire time interval that the latch is transparent. The principal advantage of latches over flip-flops is that the data is allowed to propagate through the latch as soon as it arrives instead of waiting for a clock edge.

On the other hand, timing verification tools do not yet standardly support latches, as they have more complicated timing restrictions due to e.g. time borrowing (which will be explained later on). Therefore, in the standard digital design flow, flip-flops are widely supported and flip-flop-based pipelining schemes are therefore the more established option.

4.3.3.1 Non-Overlapping Clocks

Latch-based pipelines are usually controlled by *non-overlapping* clock signals to avoid race conditions, i.e. to avoid that data can unwantedly propagate through multiple successive latches during one clock phase. The non-overlapping clocks are generated by the circuit depicted in Fig. 4.8. This Non-Overlapping Clock Generator (NOCG) generates two output signals *en_a* and *en_b* which explicitly have a delay $t_{\text{non-overlap}}$ between their high phases. This delay is determined by the delay of the NAND gate and by the delay of the inverters indicated by the gray rectangles:

$$t_{\text{non-overlap}} = t_{\text{NAND}} + n \cdot t_{\text{INV}} \tag{4.27}$$

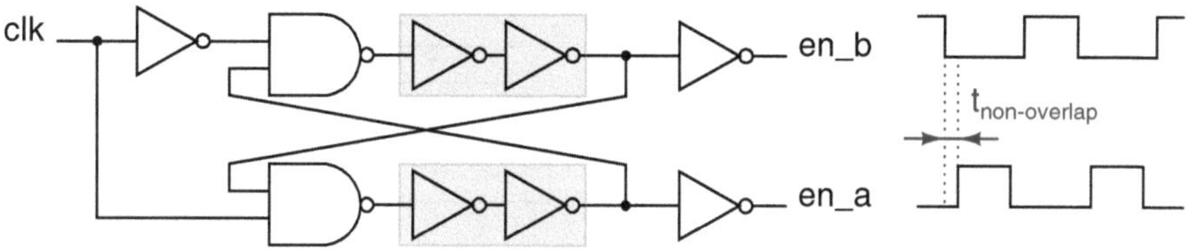

Fig. 4.8 Schematic of non-overlapping clock generator

Table 4.2 Non-overlapping clock generator implementations in the four prototypes

Prototype	Adder	MAC	MAC	JPEG
CMOS technology	90 nm	90 nm	40 nm	40 nm
# of inverters *n*	0	4	6	6

By varying this amount of inverters, the parameter n in the equation changes and the delay can be altered. As long as n remains even, the non-overlapping functionality is guaranteed. In the figure, n is equal to 2. The delay $t_{\text{non-overlap}}$ can thus be chosen at design time according to the needs of the circuit and of the technology. Note that the non-overlap time between the clock signals does not degrade performance of the latch-based system, because data continues to propagate through the combinational logic between the latches even while both clocks are low [6].

This NOCG has been used in the four prototypes, albeit with different values of parameter n, as shown in Table 4.2. The difference between the MAC design implementations is due to the increased variability in the 40 nm CMOS technology. Important to note is that the NAND gates in the NOCG are implemented in standard CMOS logic, since due to the NAND implementation with naturally stacked nMOS transistors, the sizing of the pMOS transistors is rather modest. Moreover, no differential signals are available, which makes TG logic unusable. On the other hand, excessive pMOS sizing is required for the inverters, hence, they are implemented as stacked nMOS inverters, as everywhere throughout the prototypes.

4.3.3.2 Time Borrowing

In latch-based pipelines, if the pipeline stages are not perfectly balanced, a slower stage may borrow time from a faster stage, since data can propagate as soon as it arrives in the transparent phase of the latch. Therefore, some stages can be longer while other are shorter, and the latch-based system will tend to operate at the average of the delays, while a flip-flop-based system would operate at the longest delay. This ability of slow logic in one pipeline stage to use time nominally allocated to faster logic in another stage is called *time borrowing* [1].

Figure 4.9 illustrates this concept, where the longer cycle opportunistically borrows time from the next cycle. Moreover, time borrowing may operate over multiple stages: it may continue indefinitely so long as the data never arrives so late at a latch that its setup time is violated. The setup time correlates to the time

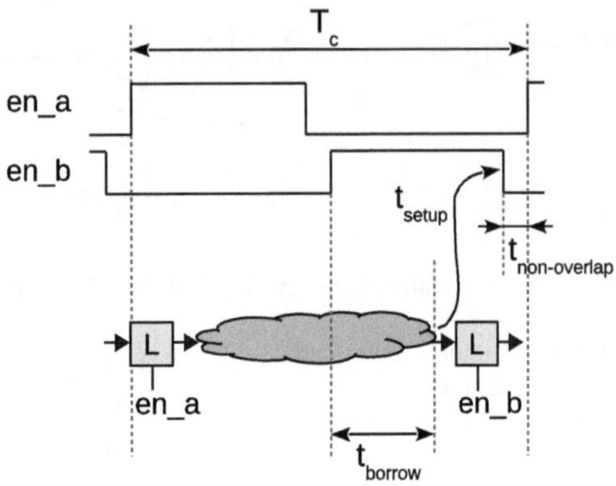

Fig. 4.9 Concept of time borrowing and its limitation

that the input data must have settled before the end of the transparent phase of the latch, i.e. the falling edge of *en_b* in this case, so that the input is correctly captured. The maximum amount of time t_{borrow} which may be borrowed is close to half a clock period T_{c} [6]:

$$t_{\text{borrow}} \leq \frac{T_{\text{c}}}{2} - \left(t_{\text{setup}} + t_{\text{non-overlap}}\right) \tag{4.28}$$

Hence, time borrowing is, aside from the clock period, limited by the setup time of the latch and by the non-overlap time, as shown in Fig. 4.9. Although the non-overlap time does not degrade the system's performance, as mentioned earlier, it does restrict the allowed amount of time borrowing.

Although time borrowing averages out the delay of an unbalanced pipeline, this does not signify that intentionally designing an unbalanced pipeline is a good idea in ultra-low-voltage designs, on the contrary. However, time borrowing does allow to opportunistically borrow time whenever variable gate delays disturb the designed balanced pipeline. Seeing the fact that timing variations of sub- and near-threshold circuits are very high, time borrowing is a very beneficial concept. The following analysis will support this claim.

To assess the effect that time borrowing has on the minimum clock period, an analysis was performed on the critical path of a latch-based versus a flip-flop-based pipeline. The test setup for this analysis is shown in Fig. 4.10. The parameters of the test setup have been based on the 90 nm MAC design, to clarify the impact of time borrowing on a specific design. Therefore, the logic depth of the pipeline stages in the critical path is 2 for the latch-based pipeline, and a pipeline depth of 32 has been employed. In order to achieve the same latency for the flip-flop-based equivalent, the same amount of logic gates has been used, while the positive latches have been replaced by flip-flops, and the negative latches by inverters. This effectively halves the number of pipeline stages for the flip-flop-based system.

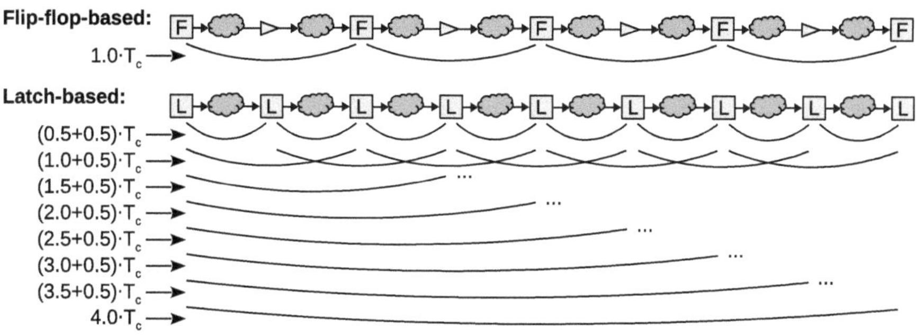

Fig. 4.10 Analysis of timing constraints on a flip-flop-based critical path versus a latch-based pipeline taking into account the effect of time borrowing, visualized for a latch-based pipeline depth of eight

The timing constraints for the *flip-flop-based pipeline* are straightforward: every stage must complete within one clock period, as visualized in Fig. 4.10:

$$t_{\text{every stage}} \leq T_c \tag{4.29}$$

Due to the time borrowing of the *latch-based pipeline*, the timing constraints become more complex. To slightly simplify the analysis, the maximum amount of time borrowing is set to be half the clock period, which is an approximation of (4.28):

$$t_{\text{borrow}} \leq \frac{T_c}{2} \tag{4.30}$$

A single pipeline stage can then use up to a clock period for completion, when its nominally allocated time and the maximum time borrowing are added:

$$\begin{aligned} t_{\text{single stage}} &\leq \frac{T_c}{2} + t_{\text{borrow}} \\ &\leq T_c \end{aligned} \tag{4.31}$$

Since time borrowing can accumulate across multiple pipeline stages, the timing constraints of multiple stages are added with the maximum amount of time borrowing ($= 0.5 \cdot T_c$) as well. In Fig. 4.10, the timing constraints for a latch-based pipeline with a pipeline depth of 8 are visualized for every combination of subsequent stages.

The latch-based critical path has to be completed within the same delay as the flip-flop-based path, in order to adequately compare both. Therefore, time borrowing cannot go beyond the last pipeline stage. The timing constraint of the last stage of a critical path of N stages thus becomes:

$$t_{\text{last stage}} \leq \frac{N}{2} \cdot T_{\text{c}} \tag{4.32}$$

These timing constraints conditions are formally described in the following equation:

$$\forall n \in \{1, 2, \cdots, N\} : \forall i \in \{0, 1, \cdots, N - n\} : \sum_{j=0}^{n-1} t_{\text{i+j}} < \frac{n+1}{2} \cdot T_{\text{c}} \tag{4.33}$$

where N is the pipeline depth, n is the length of the considered path and i indicates the starting position of that path. The equation states that the sum of all pipeline stage delays of a certain path must not be higher than $(n \cdot 0.5 + 0.5)$ times the clock period T_{c}.

Now that the formal descriptions of the timing constraints are established for both latch-based and flip-flop-based pipelines, they can be calculated using Monte Carlo (MC) simulations.

The clock period of a latch-based pipeline is determined through Eq. (4.33) by performing MC simulations of the delay of the cascaded logic and of the latch. The mean and the standard deviation of these delays are then combined using Eqs. (4.25) and (4.26) to obtain the worst-case path delays. A 6σ margin has been used, corresponding to an error probability of less than 1 out of a billion. Equation (4.33) is thus reworked to:

$$\forall n \in \{1, 2, \cdots, N\} : \left(n \cdot \mu_{\text{ps\_latch}} + 6 \cdot \sqrt{n} \cdot \sigma_{\text{ps\_latch}}\right) \cdot \frac{2}{n+1} < T_{\text{c}} \tag{4.34}$$

where $\mu_{\text{ps\_latch}}$ and $\sigma_{\text{ps\_latch}}$ are the mean delay and its standard deviation of a single pipeline stage when using latches. To conclude, the highest number of the left side of the equation sets the minimum clock period of the latch-based pipeline.

The minimum clock period of the flip-flop-based pipeline on the other hand is easily determined by calculating:

$$T_{\text{c,min\_flip-flop}} = \mu_{\text{ps\_flip-flop}} + 6 \cdot \sigma_{\text{ps\_flip-flop}} \tag{4.35}$$

Figure 4.11 shows the result of this analysis. The percentage delay gain is the relative amount by which the clock period can be reduced by using a latch-based instead of a flip-flop-based pipeline. This delay gain is due to both the cascading of gates and the time borrowing. As visible, time borrowing makes it possible to reduce the clock period by 10.6 % at the nominal supply voltage of 1 V. This makes it an interesting concept, but the real profit is obtained at ultra-low supply voltages. Here, time borrowing really stands out due to the highly increased timing variability. For example, at a supply of 200 mV, the delay gain becomes no less than 37.3 %.

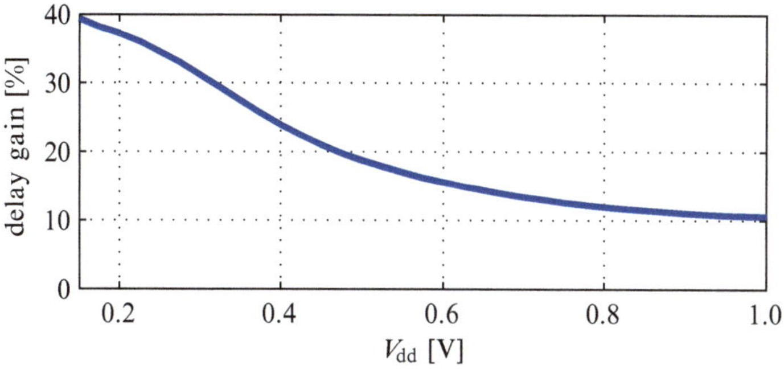

Fig. 4.11 Simulated gain in delay for a latch-based pipeline which allows time borrowing, compared to a flip-flop-based system, as function of V_{dd}

4.3.3.3 Conclusion

To conclude, variability in sub-threshold circuits results in highly variable gate delays, even when carefully balancing the pipeline. This makes a pipeline that enables time borrowing preferable in the ultra-low-voltage region. Therefore, the prototypes of this work use a pipeline based on level-sensitive latches instead of on edge-triggered flip-flops.

Another advantage of a latch-based system, is that clock skew does not degrade performance, as opposed to flip-flop-based systems [6]. Even when the clock signals are skewed, the data can still arrive at the latches while they are transparent. Therefore, latches are said to be skew-tolerant, as clock skew does not impose a performance penalty. However, it does reduce the allowed amount of time borrowing, as does the non-overlap time of the clock signals.

As mentioned earlier, unfortunately, timing verification tools do not yet standardly support latches. However, since the prototypes of this book are not designed in the standard digital design flow, this did not pose a problem at design time. More information on the employed design methodology for the prototypes will be provided in Sect. 4.4.

4.3.4 Design Considerations

To conclude the discussion on pipelining, this section summarizes the pipelining parameters which can be adjusted and the related design considerations.

Given a specified system, the two main parameters to determine are the pipeline stage length and the pipeline depth. As mentioned before, the pipeline stage length is related to the maximum logic depth. Using pipeline stage lengths beyond 1 TG logic gate implies that differential TG logic must be implemented. Counter-intuitively, constructing logic gates differentially does not increase total area or energy consumption, because it allows increasing pipeline stage length and therefore decreasing the number of latches. Since the area, leakage power and dynamic energy of a latch are significantly higher than of a TG logic gate, using differential logic results in a more area- and energy-efficient design. Moreover, it increases gate reliability and the differential implementation of the latch adds to the total variation-resilience as well.

It is advantageous to cascade as many TG logic gates as feasible, since timing variations are then averaged. Considering the large timing variations due to working in the ultra-low-voltage region, this averaging, combined with the latches' time borrowing capability, is very beneficial. The regeneration which is necessary after every cascade of TG logic is performed by latches, because they allow the previously mentioned concept of time borrowing.

However, note that it would also be possible to have a longer stage length than the maximum logic depth, by introducing inverters between cascades instead of latches. These inverters then serve as regeneration elements. The logic depth restriction is thus not equal to a pipeline stage length restriction. This allows to adjust the second parameter, the pipeline depth, according to the system's need. For example, if the goal is a predefined fixed performance, the pipeline depth and the stage length can be balanced to achieve this.

Although adding two inverters on the complementary paths is an option, this results in an almost as high leakage consumption as adding the differential latch of Fig. 3.31b which consists of the same two inverters and four transmission gates. Moreover, adding a differential latch increases the throughput of the total pipelined system. Using the maximum logic depth as pipeline stage length and inserting latches for all regeneration elements results in a *deeply pipelined* system.

In ultra-low-voltage designs, the inherently high circuit delay results in a significant contribution of leakage to the total energy consumption. By using deep pipelining with short pipeline stages, not only the throughput is enhanced, but the leakage energy is also reduced considerably due to the smaller clock period. Thereby, deep pipelining in ultra-low-voltage designs effectively shifts the Minimum-Energy Point (MEP) to a lower supply voltage [2]. In turn, this decreases the dynamic energy consumption. Both consequences of deep pipelining result in a lower absolute value of the total energy consumption compared to a non-pipelined system. Hence, the energy savings caused by deep pipelining outweigh the energy overhead of the clocked latches. E.g. measurements of the impact of deep pipelining in [2] revealed total energy savings of 30 % at 1.6 times higher performance.

To summarize, the combination of TG logic and latch-based deep pipelining is employed in the four prototypes for optimal throughput, energy-efficiency and variation-resilience.

4.4 Design Methodology

The design methodology which was employed in the prototypes of Chaps. 5 and 6 is shown in Fig. 4.12. The upper part of the figure shows the different design steps, while the lower part indicates how the layout was carried out.

4.4.1 Design

The architecture of the prototypes has first been implemented at high-level, using Matlab™ and Simulink™. Then, the different low-level building blocks (which were discussed in Chap. 3) were designed at transistor-level with the Spectre™ simulator, taking into account all possible variations. Next, these building blocks

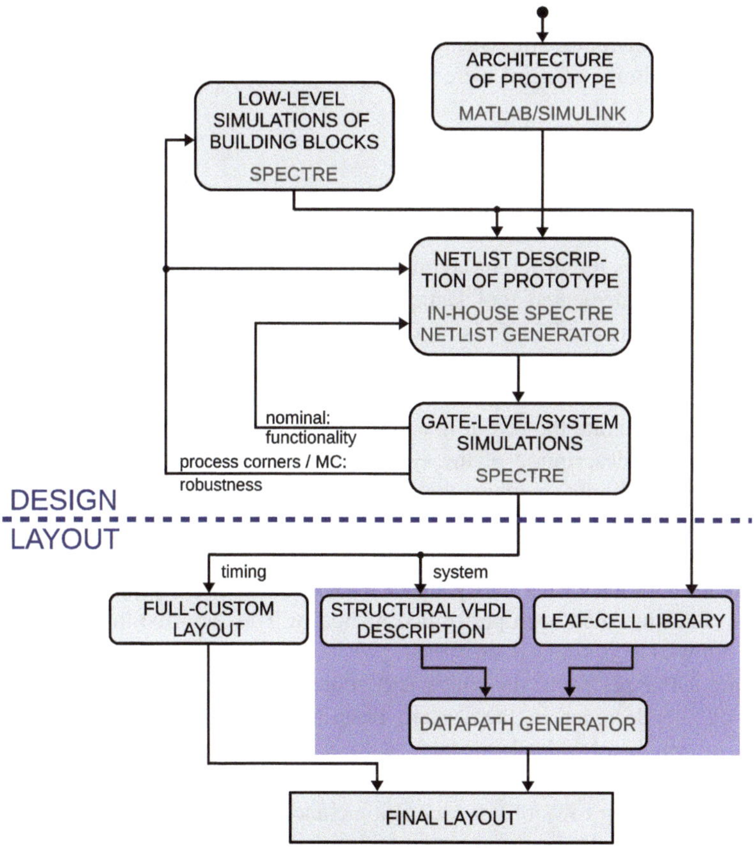

Fig. 4.12 Schematic overview of the employed design methodology

were used to construct a netlist description of the entire system with an in-house netlist generator. This in-house netlist generator uses Matlab™ code in combination with Spectre™ code to efficiently generate large netlists. Then, the system and its subblocks were exhaustively analyzed on functionality and on robustness. This analysis has also been performed with Spectre™ simulations.

Iterations were often necessary to create a variation-resilient overall design. An important side note is that circuit analysis was extensively carried out to detect which parts of the system or which subcircuits were sensitive to variations. Moreover, a sanity check was performed on all simulation results to check whether or not the result was at all reliable, considering the inaccuracy of the transistor models in the weak inversion region.

This method of designing allows to guarantee a variation-resilient design without losing the energy gain which is obtained by operating circuits in the ultra-low-voltage domain. Although sub-threshold circuits offer large energy savings, a pitfall in such design is to tackle the increased variability by taking too many design margins. In the end, the energy consumption could then be higher than what would be accomplished by working at a slightly higher supply voltage and hence less variations which require less design margins. With the proposed methodology, design and/or timing margins were only added where really necessary.

4.4.2 Layout

Figure 4.12 shows the different layout steps as well. To acquire an as dense layout as possible, the dedicated tool Datapath Generator (DPG) from RWTH Aachen University [5] was used. DPG is most useful in performing place and route of big structures that are semi-regular and thus are hardly impossible to layout manually but that can still be structurally described without too much design effort. Moreover, it is possible to iterate the layout in a very flexible and efficient manner. As inputs, DPG requires a description of the system in structural VHDL, as well as a leaf-cell library. This library contains the custom-made physical layouts of the different gate-level building blocks that are used in the prototype.

In the four prototypes, the layout of the timing blocks has been carried out full-custom, for reasons of irregularity. In the case of the JPEG encoder, the layout of the lookup tables has also been performed manually. This was due to their regularity, which allowed an optimized structure.

While DPG performs the place and route, the exact placement choices are made by the designer in the system description. This description describes a matrix in which a leaf cell is assigned to each element. Figure 4.13 shows the systematical layout structure which was used for all the prototypes. The layout contains alternating rows of (cascaded) logic and latches. The data flow goes from top to bottom and follows the architectural block diagrams of the prototypes entirely, ensuring minimal wire lengths for all signals. The layout of the timing and the clock buffers is typically positioned at the left side of the block.

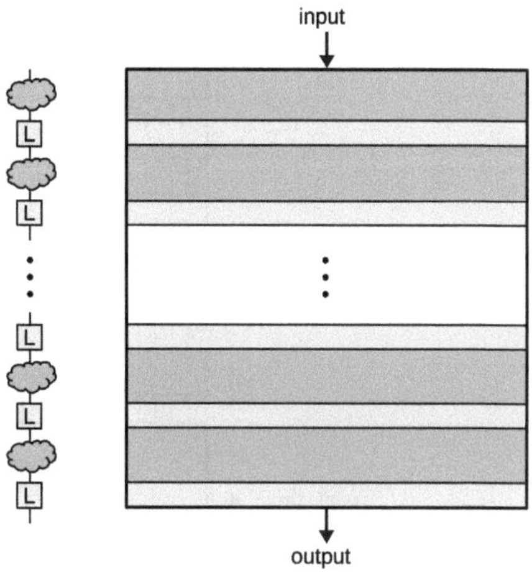

Fig. 4.13 Layout structure used with DPG in the four prototypes

The custom layout of the leaf cells can be compared to the layout of standard cells. Supply wires are connected by abutment between the different leaf cells. In case of the latches, the same is true for the clock signals. Since the latches required more area than the logic gates, their layout has been carried out first, to optimize the width of such a latch leaf cell. The idea is to minimize the width required for the latches to reduce the total horizontal wire length of the clock signals as much as possible because it is important that all latches of a certain pipeline stage receive the same clock with minimal clock skew. Taking into account the resulting width constraint for the logic gates, their leaf cells were designed to acquire minimal height, so as to minimize data wire length.

4.5 I/O Circuits

This section describes the measurement setup used for the four prototypes. I/O circuits are added at the in- and outputs of the ultra-low-voltage system for two reasons. First, they form the connection between the ultra-low-voltage system and the outside world. Standard measurement equipment is not able to interpret signals with such low voltage swing. The I/O circuits operate at a higher I/O supply, and therefore facilitate communication with the measurement equipment. The second task of this I/O circuitry is to allow at-speed functionality tests. In addition, these at-speed test circuits also enable realistic energy measurements.

Figure 4.14 shows the different blocks of the I/O circuitry and their connections with both the ultra-low-voltage system and the outside world. The shift registers can function in serial or parallel mode, as indicated by the *flag_parallel* signal. In the prototypes, the shift registers can store 8 different sets of in- and outputs. In cooperation with the ultra-low-voltage system they function as follows. First, the input data (coming from in-house developed software) is shifted serially into

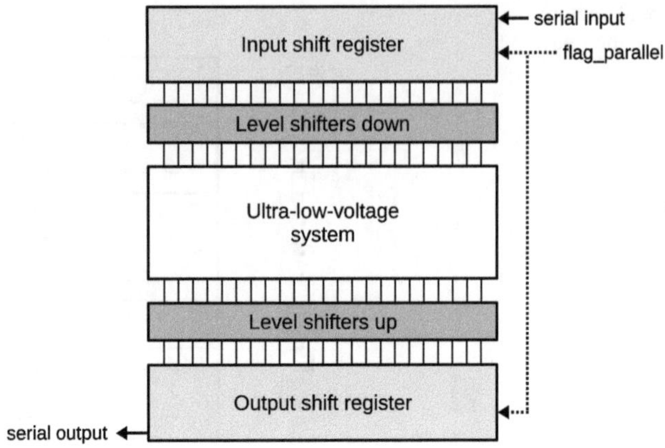

Fig. 4.14 Layout structure of connection between ultra-low-voltage system and I/O circuits used in the four prototypes

the input shift register (*flag_parallel* = 0). This input data consists of 8 separate input data sets. Then, the input and output shift registers are put into parallel mode by switching the *flag_parallel* signal to 1. The 8 sets of data flow from the input shift register through the ultra-low-voltage system. Its outputs are then stored into the output shift register. Note that the parallel clock of the shift registers is a level-shifted version of the clock of the ultra-low-voltage system, so that they operate at the same rate. Afterwards, the *flag_parallel* signal is switched again, and the output shift register is serially read out and interpreted by the software.

Since the shift registers are operating at the I/O supply, while the ultra-low-voltage system is operating at a much lower supply voltage, level shifting between these voltage levels is required. The speed of the level shifters is of crucial importance, as they have to be as fast as the system. Moreover, they are required to be very robust so that they are functional under all possible variations. The level shifters at the input need to shift the level down from the I/O supply levels $V_{\mathrm{dd,I/O}}$ and $V_{\mathrm{ss,I/O}}$ to the ultra-low-voltage supply levels V_{dd} and V_{ss}. This can be easily performed by inverters which are supplied with the ultra-low-voltage supplies, as shown in Fig. 4.15a.

The level shifters at the output are more complicated, as they need to boost the level up. As the system itself is working at extremely low supply voltages, these level shifters have to be able to sense a very small input swing and convert it correctly to the high I/O output swing. Figure 4.15b shows the schematic of the output level shifter. Regarding the sizing of the first block, the sizing of the nMOS transistor with $\overline{in}$ as input signal is crucial to determine the high level of the *int* signal. In fact, the other transistors can all have the same sizing P_1, while P_2 should be chosen so that the output swing of *int* is as large as possible. The inverter at the end should then be sized such that its switching point V_{M} matches the mean of the output swing of *int*, to guarantee as correct and robust interpretation of the output as possible under variations.

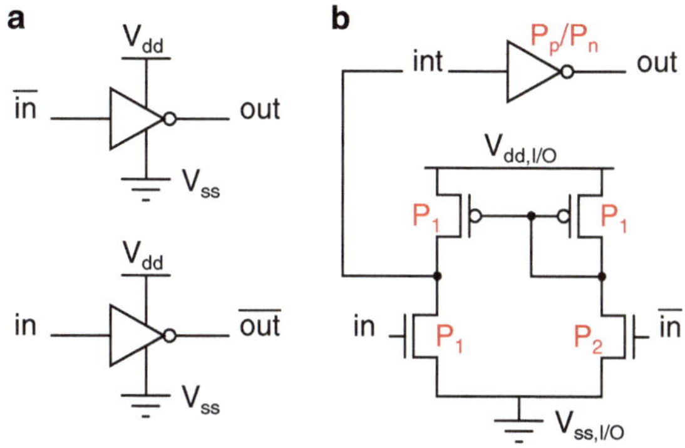

Fig. 4.15 Schematics of (**a**) level shifter down and (**b**) level shifter up

This combination of I/O circuitry will be successfully employed in the four prototypes. It not only enables functionality tests of the ultra-low-voltage systems at-speed with any arbitrary set of inputs, but it allows to perform realistic energy measurements as well because the input shift register is cyclic. In parallel mode, the inputs that are shifted out are shifted back in at the top. Hence, if the input shift register is accessed more than eight times, the data is reused.

4.6 Conclusion

This chapter reached various conclusions on the architectural design of ultra-low-voltage systems. Firstly, the energy consumption of a system has been studied from a theoretical point of view. In the subsequent chapters, which will present the design and the measurement results of the prototypes of this book, these theoretical findings can be validated by experimental data.

Secondly, it was found that cascading logic gates has a number of advantages, especially seeing the high sensitivity to variations in the ultra-low-voltage region. The resulting averaging of variations shows promising results, especially when cascading differential TG logic. This will be further investigated and analyzed in Chap. 5.

Thirdly, this chapter examined pipelined architectures and came to the conclusion of favoring latch-based pipelining over flip-flop-based pipelining for systems operating at ultra-low supply voltages, because it allows time borrowing. This is a very beneficial concept considering the high timing variability, as the simulation analysis showed. In order to avoid race conditions, non-overlapping clock signals shall be used in all the prototypes of this book. This chapter elaborated as well on the benefits of deep pipelining for ultra-low-voltage designs.

Fourthly, the design methodology which is employed for the prototypes of Chaps. 5 and 6 has been extensively presented.

Finally, the I/O circuits which are placed around the ultra-low-voltage prototypes to enable communication with the outside world and to perform measurements have been discussed.

References

1. Harris D (2001) Skew-tolerant circuit design. Morgan Kaufmann, San Francisco
2. Jeon D, Seok M, Chakrabarti C, Blaauw D, Sylvester D (2012) A super-pipelined energy efficient subthreshold 240 MS/s FFT core in 65 nm CMOS. IEEE J Solid State Circuits 47(1):23–34. DOI: 10.1109/JSSC.2011.2169311
3. Reynders N, Dehaene W (2012) Variation-resilient building blocks for ultra-low-energy sub-threshold design. IEEE Trans Circuits Syst Express Briefs 59(12):898–902. DOI: 10.1109/TCSII.2012.2231022
4. Reynders N, Dehaene W (2012) Variation-resilient sub-threshold circuit solutions for ultra-low-power digital signal processors with 10 MHz clock frequency. In: Proceedings of the IEEE European solid-state circuits conference (ESSCIRC), pp 474–477. DOI: 10.1109/ESSCIRC.2012.6341358
5. Weiss O, Gansen M, Noll T (2001) A flexible datapath generator for physical oriented design. In: Proceedings of the IEEE European solid-state circuits conference (ESSCIRC), pp 393–396
6. Weste N, Harris D (2011) CMOS VLSI design: a circuits and systems perspective, 4th edn. Addison-Wesley, Boston

Chapter 5
Datapath Blocks

This chapter presents the design of the first three ultra-low-voltage prototypes which have been implemented in this book. These prototypes all consist of datapath blocks. Their target was to be able to operate at ultra-low supply voltages, while achieving high energy-efficiency, a speed of $n \times 10\,\text{MHz}$ and a high yield through variation-resilience. This chapter builds further upon the conclusions of the analyses of different gate-level building blocks in Chap. 3 and of architectural design choices in Chap. 4.

Section 5.1 discusses the design of the first prototype, which is a 32-bit logarithmic adder fabricated in a 90 nm CMOS technology [8]. This prototype is employed as a proof of concept to confirm the robust operation of TG logic and latch-based deep pipelining in the ultra-low-voltage region. Extensive measurement results evaluate their successful functionality and are compared to the state-of-the-art.

Section 5.2 examines the ultra-low-voltage design of a 16-bit Multiply-Accumulate Unit (MAC) [9]. This MAC has been fabricated in both the 90 nm and the 40 nm CMOS technologies at hand, resulting in the second and the third prototype. Gate-level and architectural improvements with respect to the design of the adder are implemented and tested. An extensive comparison between the measurement results of the MAC in both CMOS technologies allows studying the impact of scaling on ultra-low-voltage designs [10], and to extend the scaling analysis provided in Chap. 2.

N. Reynders, W. Dehaene, *Ultra-Low-Voltage Design of Energy-Efficient Digital Circuits*, Analog Circuits and Signal Processing,
DOI 10.1007/978-3-319-16136-5_5

5.1 Adder

5.1.1 Proof of Concept

The first ultra-low-voltage prototype which has been implemented in this book is a 32-bit adder [8]. It has been fabricated in a 90 nm CMOS technology.

The research targets of this proof of concept are the following. Firstly, the chosen circuit topologies of Sect. 3.2, i.e. TG logic and standard CMOS inverters with nMOS stacking, are optimized for ultra-low-voltage operation and tested to confirm their functionality. Secondly, the proposed latch-based deeply pipelined architecture from Sect. 4.3 is implemented in the architecture of the adder and is also evaluated during the measurements. Thirdly, an on-chip power gating scheme is employed to reduce leakage in standby mode.

The aim of this design is to achieve very low energy consumption at MHz-speed, while guaranteeing variation-resilience. The following sections will discuss the architecture of the adder, the design for ultra-low-voltage operation, the measurement results and will provide a state-of-the-art comparison.

5.1.2 Architecture

For addition of numbers with high number of bits, *tree* adders are a good option since they make a trade-off between delay and energy or area. They are also called *logarithmic* adders, since their delay is proportional to $\log_2(N)$, with N the amount of bits.

Common logarithmic adder topologies are the Brent-Kung [3] and Kogge-Stone [7] topologies. Brent-Kung (BK) is characterized by its large number of logic levels, while Kogge-Stone (KS) has a high number of wires. For this 32-bit adder prototype, a Han-Carlson topology has been preferred, since it is a combination of the previously mentioned topologies and therefore makes a compromise between the number of pipeline stages and the number of wires. Figure 5.1 shows the block diagram of the Han-Carlson adder. As can be seen, its first and last stage are BK stages, while the intermediate stages come from KS.

Logarithmic adders make use of so-called generate and propagate logic. The implementation of the different elements of the adder is shown in Fig. 5.2.

5.1.3 Ultra-Low-Voltage Design

This section provides the design details of the ultra-low-voltage adder, with respect to the building blocks and architectural trade-offs which have been discussed in the previous chapters.

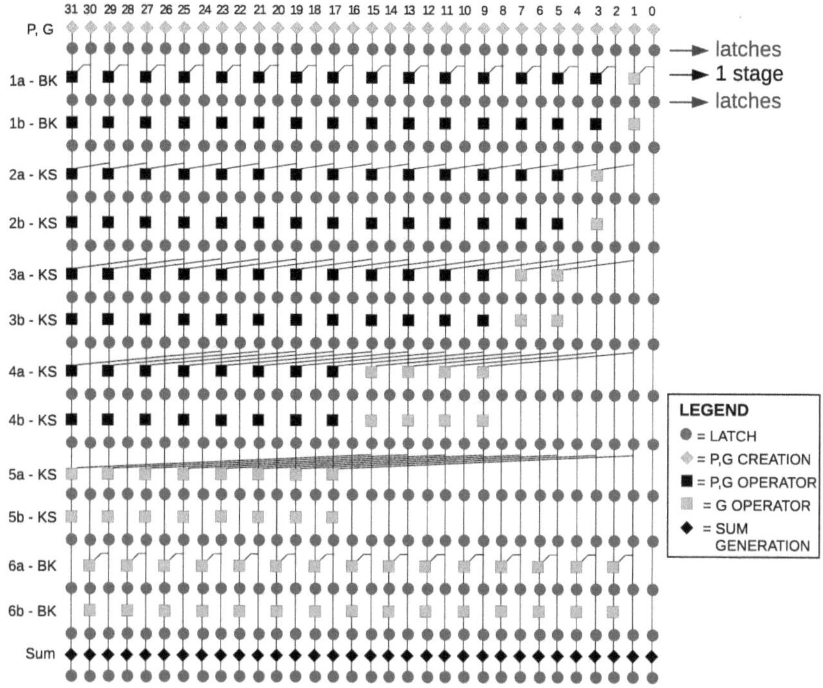

Fig. 5.1 Block diagram of the 32-bit Han-Carlson adder [8]

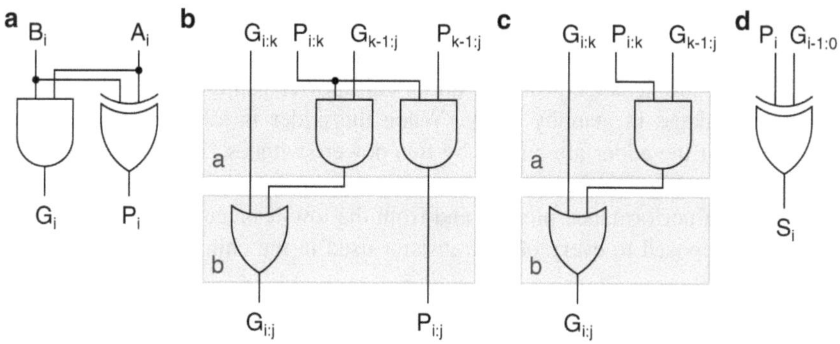

Fig. 5.2 Implementation of adder elements: (**a**) P,G creation, (**b**) P,G operator, (**c**) G operator and (**d**) sum generation

5.1.3.1 Architectural Design and Timing

In the design of the adder, pipeline stages with minimal length are employed, i.e. each pipeline stage is composed of a single logic gate, and latches are inserted after every logic gate. As can be seen in Fig. 5.2, the P,G creation and sum generation elements require only one logic gate. Hence, these elements are implemented without difficulty in the first and last pipeline stages of the adder (see Fig. 5.1). However, each P,G operator and G operator of the adder requires two consecutive logic gates. Those operators are therefore split in two parts *a* and *b*, thereby explaining the stage descriptions of the intermediate stages of the adder in Fig. 5.1.

The latch-based pipeline is controlled by non-overlapping clock signals to avoid race conditions (refer also to Sect. 4.3.3 for an in-depth explanation). Because of the very short pipeline stages, the adder is maximally pipelined, which should increase the throughput significantly. Such deep pipelining also increases latency, but the state-of-the-art comparison in Sect. 5.1.5 will show that the latency of this design is still much lower than the one of the comparison point.

5.1.3.2 Gate-Level Building Blocks

Transmission Gate (TG) logic is used for the design of the logic gates. As visible in Fig. 5.2, only a few different logic gates are required: AND, XOR and OR. Single-ended TG logic is implemented, with a latch after every TG logic gate. Since no differential input signals are available, the single-input latch (Fig. 3.31a) of Chap. 3 is employed. The latches provide differential output signals, which serve as inputs for the TG logic. The sizing of the building blocks can be found in Sect. 3.4.

5.1.3.3 Power Gating

As the supply voltage lowers and the circuit operates more in the sub-threshold region, leakage power becomes an important part of overall power consumption. To overcome the idle leakage problem, an on-chip power gating scheme is employed to reduce leakage in standby mode. When the adder is idle, the power supply and ground of the adder are cut off by two power switches. Their implementation can be seen in Fig. 5.3. In this 90 nm technology, it is possible to use transistors from the high-performance process and from the low-leakage process on the same design. As opposed to every other transistor used in the chip, transistors from the

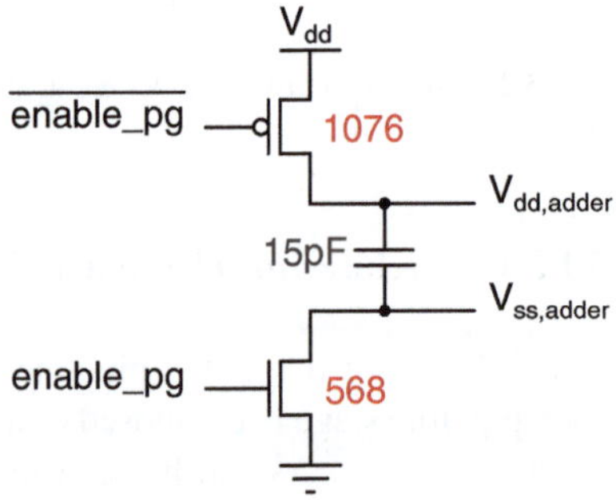

Fig. 5.3 On-chip power gating scheme and decoupling capacitor of the adder

low-leakage process are used for the power gating, since these transistors allow a much higher leakage reduction than transistors in the high-performance process of the technology.

The sizing of the transistors in Fig. 5.3 is such that they can deliver the required current necessary for the adder. The decoupling capacitance value is determined so that all possible peak currents are managed.

5.1.4 Measurement Results

Figure 5.4 shows the chip micrograph of the 32-bit Han-Carlson adder, implemented in a 90 nm CMOS technology. The active area of the adder is $650 \times 50\,\mu\text{m}^2$.

Measurement results of the adder as function of the supply voltage are given in Fig. 5.5. A total of eight dies has been measured. All plots show the mean, maximum and minimum values which were obtained from the different dies. All values mentioned in the text are mean values.

The upper plot shows the obtained maximum clock frequencies per supply voltage. The adder is fully functional down to a supply of 190 mV, thereby confirming that the building blocks are operational in the sub-threshold region. This low $V_{\text{dd,min}}$ also proves that the design is variation-resilient, as otherwise it would not be functional at such low supply voltages. At the minimal V_{dd}, the clock frequency is 10 MHz.

The middle plot provides the measured energy consumption per operation when the adder is operating at the clock frequencies of the upper plot. At the minimal supply, the energy consumption per operation is 0.4 pJ at a 10 MHz clock frequency. A performance of 30 MHz is obtained at a supply of 260 mV and an energy consumption of 0.60 pJ per addition. Operational frequencies in the 10 to 100 MHz range are achieved at an energy consumption per addition that is below 1 pJ.

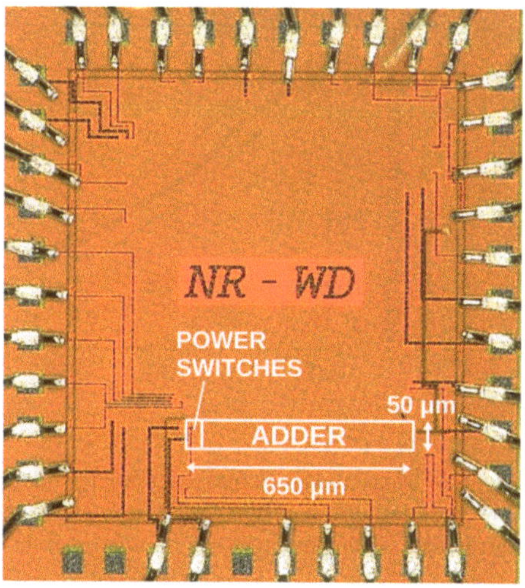

Fig. 5.4 Chip micrograph of the 32-bit Han-Carlson adder

The energy consumption seems to increase approximately linearly with V_{dd}, while the clock frequency increases roughly quadratically. An application requiring higher performance can thus achieve this at a relatively low energy cost, which demonstrates the versatility of this design.

The lower plot of Fig. 5.5 shows the measured contribution of the leakage power to the total power consumption of the 32-bit adder as function of V_{dd}. The total power consumption is measured during operation at the maximum operating frequency for every supply voltage. The leakage power is measured when the adder is idle. At a supply of 190 mV, the leakage power is 1.65 μW and leakage consumes

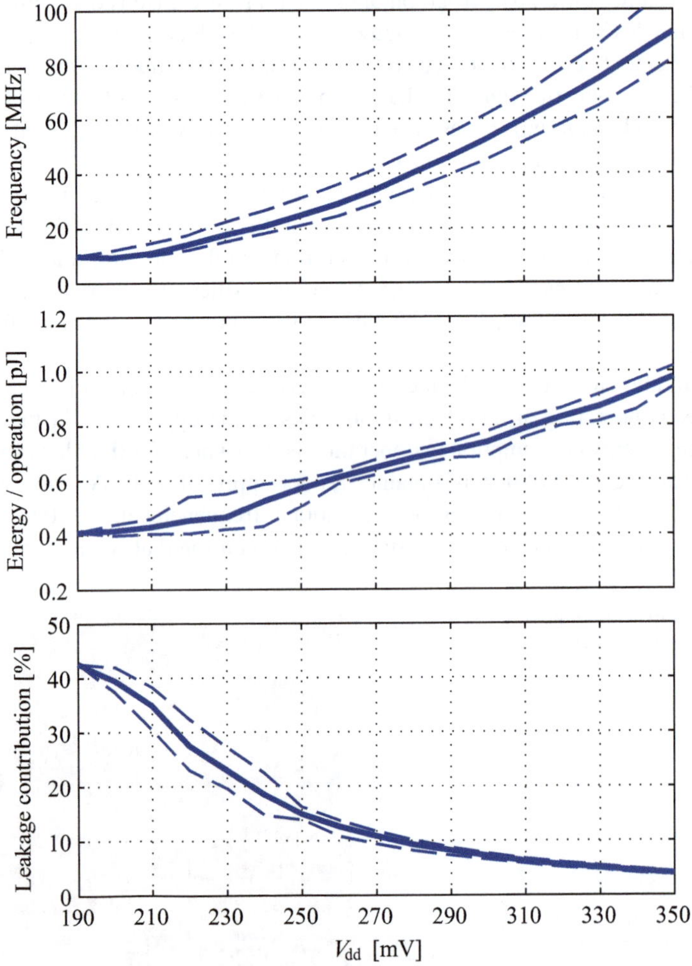

Fig. 5.5 Measured clock frequency, energy consumption per operation and contribution of leakage power to the total power consumption, as function of V_{dd}. Mean values are indicated in (*bold*), surrounded by the maximal and minimal measured values out of eight dies

42.6 % of the total power consumption. As can be seen in Fig. 5.5, the leakage contribution becomes very high when going to ultra-low supply voltages.

The on-chip power gating scheme results in a measured standby leakage reduction of approximately a factor 10, thereby proving its effectiveness.

5.1.5 *State-of-the-Art Comparison*

Table 5.1 compares this adder with the only other published sub-threshold adder. It consists of a 32-bit Kogge-Stone adder [4], also fabricated in a 90 nm CMOS technology. Seeing the similarities of both designs, they can be easily compared. The table provides a comparison on throughput (T), latency (L), energy consumption per operation and Energy-Delay Product (EDP). The EDP is the FOM that balances the importance of both energy and performance.

As can be seen, this work outperforms the Kogge-Stone adder in both throughput and latency at ultra-low supply voltages. While [4] operates in the kHz-range in the sub-threshold region, this Han-Carlson adder functions in the MHz-range. The referenced adder [4] focused on minimum energy operation and obtains a lower energy consumption, while the adder design of this book achieves the combination of both ultra-low-energy and MHz-performance, as is obvious from the improvement of the EDP, which is 160 to 900 times lower than the EDP of [4]. At much higher supply voltages, [4] does reach MHz-performance, i.e. frequencies above 10 MHz are achieved starting from a V_{dd} slightly below 600 mV at an energy consumption around 0.25 pJ. This shows that for the next prototypes in this text it would be recommended to decrease the amount of latches compared to the amount of logic gates in order to reduce the total energy consumption.

Table 5.1 State-of-the-art ultra-low-voltage adder comparison [8]: both works are 32-bit logarithmic adders in 90 nm CMOS technologies

	[4]: Kogge-Stone				This work: Han-Carlson				
V_{dd} [mV]	T [Hz]	L [μs]	Energy [pJ]	EDP [pJ.μs]	T [Hz]	L [μs]	Energy [pJ]	EDP [pJ.μs]	**EDP Factor**
190	–	–	–	–	9.5 M	1.47	0.408	0.043	–
250	7 k	143	0.145	20.7	24.8 M	0.57	0.569	0.023	**900**
300	25 k	40	0.100	4.0	52.5 M	0.27	0.739	0.014	**285**
330	50 k	20	0.095	1.9	74.6 M	0.19	0.868	0.012	**160**

5.1.6 *Conclusion*

This section described the design of a 32-bit Han-Carlson adder in a 90 nm CMOS technology. The measurement results of the adder have allowed to successfully validate the gate-level building blocks and the architectural design choices of this book for ultra-low-voltage operation. The employed techniques enabled operation down to a supply of 190 mV. The measurements demonstrate that it is possible to achieve MHz-speed combined with sub-pJ energy consumption. This work achieves a significant improvement in EDP of up to a factor 900 compared to the state-of-the-art in sub-threshold adder design. A simple on-chip power gating scheme was effective to reduce standby leakage power.

The results of this adder were very promising and have been used as a basis to build further upon for later designs, as will be discussed in the following section.

5.2 Multiply-Accumulate Unit

5.2.1 *Proof of Concept*

The second ultra-low-voltage prototype of this book consists of a 16-bit MAC [9]. It has been fabricated in the same 90 nm CMOS technology as the first prototype. The MAC is considered as the basic building block of many DSP algorithms and is thus very frequently used in Digital Signal Processor (DSP) designs. To illustrate the efficacy of the proposed design strategy in this book, the MAC has been chosen as a test case to prove that this strategy can be applied to all DSP blocks operating in the ultra-low-voltage region. Moreover, it is a complex block that includes feedback.

Another aim of the 90 nm MAC has been to optimize some gate-level and architectural design choices after measurements of the first prototype, i.e. the adder, in the same 90 nm technology. Since the pipeline stage length is only 1 in the adder, the clock frequency of the system is determined by the worst-case gate in terms of timing. Because the timing variations are high in the ultra-low-voltage region, the clock frequency is really dominated by those variations. Moreover, a latch was inserted after every TG logic gate, which results in a large amount of latches. As explained before, the area, leakage power and dynamic energy of those latches are significantly higher than those of a TG logic gate.

Therefore, the decision was taken to increase the stage length which results in averaging of timing variations of consecutive gates (as explained in detail in Sect. 4.2). The worst-case timing of a pipeline stage then still determines the clock frequency, but the impact of variations is reduced considerably due to averaging. Additionally, increasing the stage length decreases the number of latches which reduces the area and improves the energy consumption.

The third prototype has been implemented to study the effect of technology scaling on ultra-low-voltage circuits. Therefore, the 16-bit MAC has been redesigned and fabricated in a 40 nm CMOS technology [10]. Until now, the impact of CMOS technology scaling on circuits operating in the ultra-low-voltage region has received little attention, limited to device-level studies and circuit-level simulations, as discussed in Sect. 2.3. To fill the hiatus between simulations and measured, confirmed results, the MAC has been designed, processed and measured in both a 90 nm and a 40 nm CMOS technology.

The following sections will discuss the architecture of the MAC, the ultra-low-voltage design choices, the measurement results and a state-of-the-art comparison.

5.2.2 *Architecture*

The implemented chip is not only able to perform Multiply-Accumulate operation, but the system is expanded so that it can also operate in two other operation modes: multiplier (MULT) and multiply-add (MADD) mode. The equation that represents the operation of the system is $out = A.B + C$ where A and B are N-bit binary numbers and C is a 2N-bit number. The output *out* must be at least a 2N-bit number. The usage of C changes for each operation:

- MAC operation: the input C is in fact the previous output out_{prev}.
- MULT operation: C is changed to zero.
- MADD operation: C represents a third, 2N-bit input.

Standard MAC designs consist of a multiplier followed by an adder to perform the accumulation, as visualized in Fig. 5.6. However, deep pipelining is not possible with such a standard design, because the maximal pipeline depth is only 2. To allow more pipelining, the structure of the MAC implementation has to be changed. Since the basic form of multiplication can be reduced to the addition of partial products, extending this addition step replaces the need for the separate accumulation step. This can be seen in Fig. 5.7. The implemented multiplier is based on the Modified Baugh-Wooley multiplier algorithm [5], which is an algorithm that allows efficient multiplication of signed numbers.

This design thus consists of a multiplier which is extended with an interwoven accumulation structure. Figure 5.8 shows a functional diagram of the implemented MAC where the interwoven diagonal accumulation is clearly visible. This accumulation is obtained through 32 dedicated feedback (FB) latches which perform

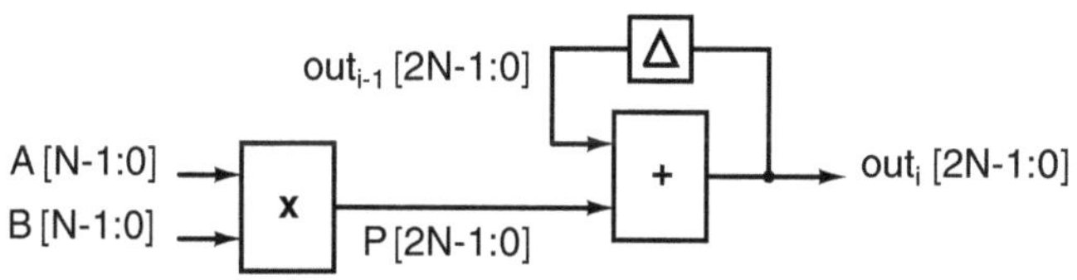

Fig. 5.6 Standard MAC implementation [10]

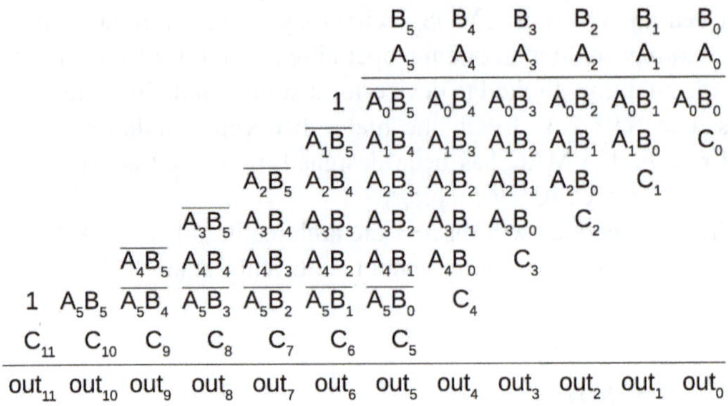

Fig. 5.7 Modified Baugh-Wooley algorithm for 6-bit signed multiplication (in *black*), extended with accumulation (in *gray*)

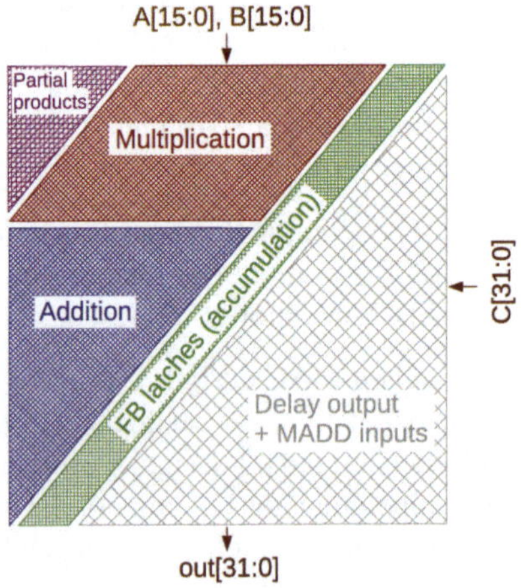

Fig. 5.8 Architecture of the 16-bit Multiply-accumulate unit [10]

bit-by-bit feedback of the previous output. Not only does such an interwoven implementation allow to efficiently pipeline the architecture, it also significantly reduces the total delay for the multiply-accumulate operation to a delay slightly higher than needed for multiplication only. Another advantage is the much higher throughput that can be achieved through deep pipelining.

Figure 5.9 gives the detailed gate-level architecture of the MAC. The MAC is able to work with two's complement numbers. The implemented MAC operation consists of 16-bit multiplication with 32-bit accumulation. On system level, the choice of operation mode is performed through two configuration bits that direct three timing control signals. Those timing signals are fed to the feedback latches which therefore enable the operation in the different modes. Calculations in every mode take the same amount of latency and there is no delay penalty for a multiply-add or multiply-accumulate operation with respect to multiplication.

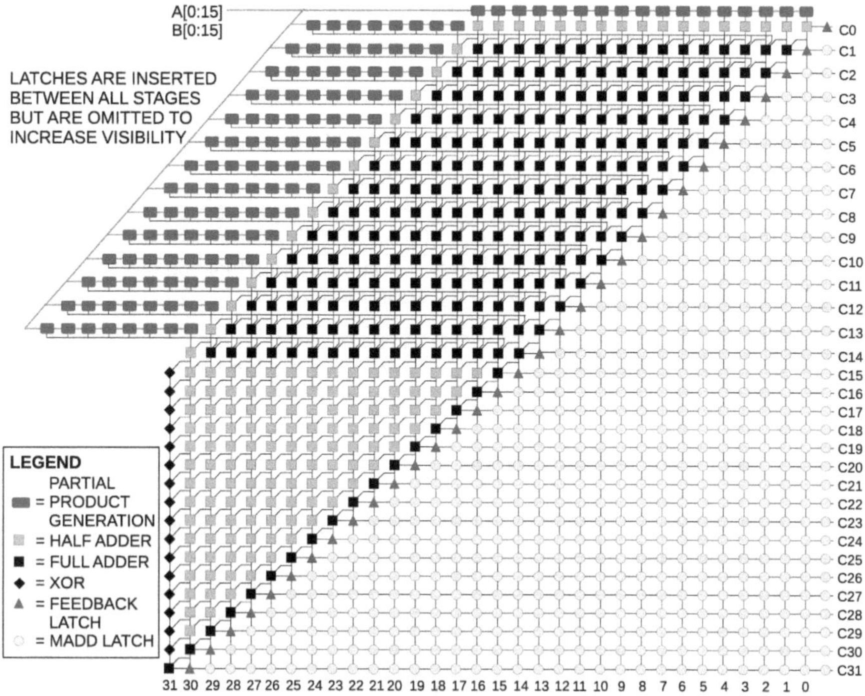

Fig. 5.9 Block diagram of the 16-bit Multiply-accumulate unit [9, 10]

5.2.3 *Ultra-Low-Voltage Design*

This section explains in detail the architectural design and transistor-level implementation of the different logic components. The differences in implementation between the 90 nm and 40 nm CMOS technologies are also addressed. In addition, the implementation of the timing is described, with a specific focus on the different design decisions necessary for both technologies. Note that in both technologies, all transistors used in the chips are LVT devices, to be able to make a fair comparison.

5.2.3.1 Architectural Design

Increasing the pipeline stage length has the attractive properties of averaging of variations and reducing the number of latches. However, TG logic requires differential input signals. These could be provided by introducing inverters, but this would result in a very large cost in leakage power and energy consumption, and would counteract the reduced amount of latches. Another possibility is to implement all logic gates differentially, which solves the issue and costs much less in terms of area and energy, in this specific configuration. Moreover, differential implementation of logic improves the gate reliability and the variations on the timing, as discussed in Sect. 4.2.3.

There are limitations to increasing the pipeline stage length as well: cascading too many TG logic gates causes a signal loss which could compromise robust behavior. Therefore, the maximum stage length is determined by whether or not the latch can still interpret its input correctly, taking into account variations. In the 90 nm technology, the maximum logic depth is determined to 2 for the target minimum supply of the design, i.e. 150 mV. The same pipeline stage length is used in the 40 nm version of the MAC, for comparison purposes. To conclude, although the pipeline stage length is doubled with respect to the adder, the architecture of the MAC is still deeply pipelined.

5.2.3.2 Gate-Level Building Blocks

Differential TG logic has been used for the **logic gates** in the MAC designs, which significantly increases the variation-resilience. The sizing of the TG logic in both technologies has been provided in Sect. 3.4. Compared to the adder, the width of the pMOS transistors is doubled because the variations in voltage level of the output of cascaded logic gates revealed to be much larger for high level than for low level. Increasing W_{pMOS} aids considerably to limit these variations on the logic high level. Moreover, it increases the problematic $I_{\mathrm{on,pMOS}}/I_{\mathrm{off,nMOS}}$ ratio, which is beneficial.

The main logic elements for the MAC are the half and full adders. A half adder (HA) can be easily implemented within the stage length boundaries (see Fig. 5.10a), but for a full adder (FA), this is a challenge. When only using logic gates with two inputs, the minimal logic gate depth of a FA is 3. Fortunately, one 3-input logic gate is possible with a single TG logic block: a multiplexer. Therefore, it was possible to satisfy the pipeline stage length boundary of 2 with the implementation of the FA, as visible in Fig. 5.10b.

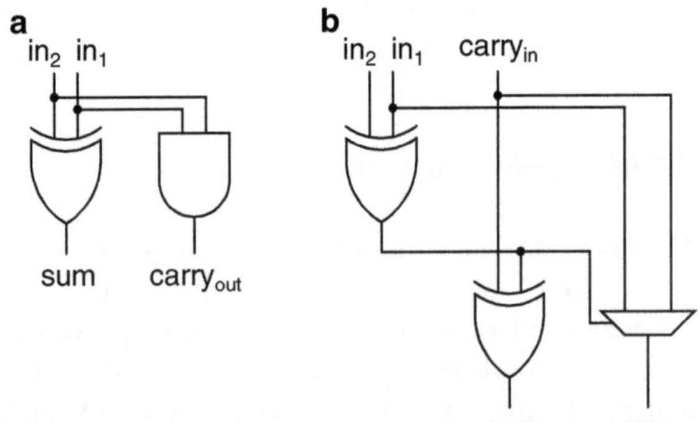

Fig. 5.10 Implementation of MAC elements: (**a**) half adder and (**b**) full adder

The **latch** that has been employed in the MAC is a fully differential latch, see Fig. 3.31b for the implementation. This is possible because complementary input signals are available from the differential TG logic. Due to the differential nature of the latch, it exhibits improved robustness compared to a single-input latch, which adds to the variation-resilience of the total design.

The transistor sizing of the latch is available in Sect. 3.4. The sizing of the inverters was not chosen minimally for several reasons. In both technologies, MC simulations showed that in some cases the output signal was not stable even though the latch was locked. This can be explained by unwanted leakage paths through the logic gates which were connected to the output of the latch and by the intra-die variations which severely weakened one of the inverters in the cross-coupled part of the latch. Normally, such cross-coupled inverters are advantageous because they regenerate the input signals and really pull their levels to the supply rails, whereas a single inverter would simply amplify the output signals. However, when one of these cross-coupled inverters becomes too weak due to variations, the feedback in the loop actually accelerates an unwanted bit flip. A solution for this problem is upsizing the inverter, which increases the drive strength of the inverter and reduces its sensitivity to variations.

Another reason was that it is imperative that the latch always interprets its input signals correctly, under all possible variations. Upsizing helps again in this case because of the decreased variability. The amount of upsizing was then determined by calculating the probability of failure of a latch under variations: the distribution of the output level of a signal that was propagated through a chain of TGs was compared to the distribution of the offset voltage of the latch:

$$P(\text{failure}) = \int_x P(\text{level}_{\text{out}} = x) \cdot P(V_{\text{offset}} > x) \cdot \mathrm{d}x \tag{5.1}$$

In the 90 nm technology, upsizing with a factor 1.5 proved to be sufficient, but due to the increased variability, the 40 nm version needed a slightly higher factor of 2.

As explained in Sect. 5.2.2 on the MAC architecture, the **feedback latches**, which are placed on the main diagonal of the MAC, enable operation in three different modes of the MAC. Figure 5.11 shows the schematic of the FB latch. The elements of the FB latch are identical to the ones of the regular latch, except for the reset transistors. The reset is not performed by a single transistor at one side of the cross-coupled inverters because this would lead to ratioed design, which is to be avoided in sub-threshold circuits due to their high sensitivity to variations. Therefore, the reset has to be performed at the two sides of the inverters without cross-coupling them through the TGs.

A differential reset requires a pull-up transistor on one side and a pull-down transistor on the other side, as can be seen in Fig. 5.11. An important consideration is the leakage contribution of the reset mechanism, since the storage functionality of the latch can be disturbed by this leakage. A minimal pMOS is therefore chosen at both sides to reduce leakage, because such a minimal pMOS leaks significantly

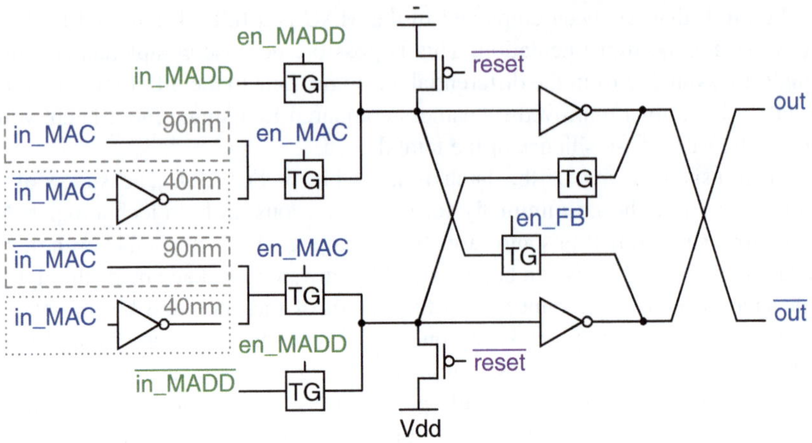

Fig. 5.11 Schematic of the feedback latch [10]

Fig. 5.12 Timing diagram of the different operation modes of the feedback latch [10]

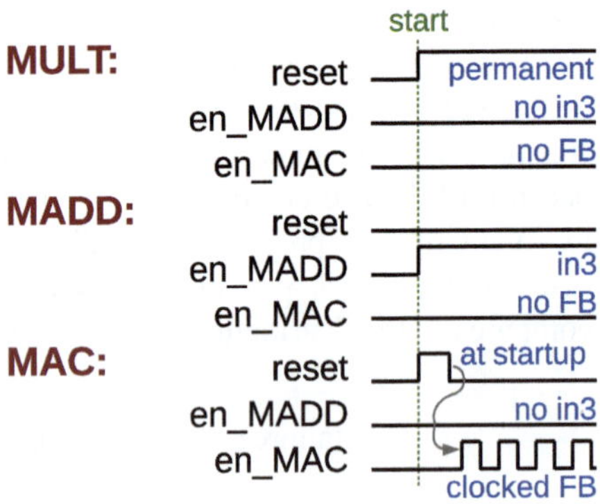

less than a minimal nMOS transistor in both technologies (e.g. 12.6 times less at $V_{dd} = 150\,\text{mV}$ in the 90 nm technology). Moreover, using two pMOS transistors requires only a single reset signal, instead of two complementary signals, which facilitates routing.

The timing configuration per operation mode of the FB latch can be seen in Fig. 5.12:

- MAC mode: The clock signal *en_MAC* is configured to be complementary to *en_FB*. The MAC input bits come from the previously calculated product out_{prev} used for accumulation and the MADD inputs bits are cut off (*en_MADD*=0). At startup of the MAC mode, it is compulsory that the FB latches are reset to ensure the first accumulation with zeros.
- MULT mode: The FB latches are permanently reset while all input bits are cut off (*en_MAC*=0 and *en_MADD*=0), so that addition with 0 is ensured. The reset mechanism is slightly complicated to remove any ratioed design: at startup, the reset is pulled high, while *en_FB* is kept low. After a certain amount of time

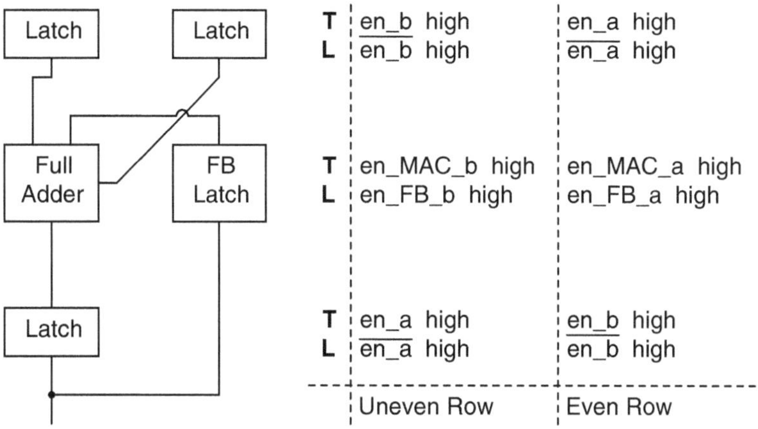

Fig. 5.13 Zoomed in part of the diagonal accumulation of the MAC, with the timing signals of the throughput (T) and locked (L) phases of the latches and the feedback latch added according to the row [10]

when it is sure that the reset node signal levels have settled, *en_FB* is pulled high to establish the regeneration characteristic of the cross-coupled inverters and ensure that signal levels are full-swing.

- MADD mode: Feedback is cut off (*en_MAC*=0) and a third 32-bit input *C* is provided from the right side, as visible in Fig. 5.9. Due to the pipelining, these MADD input bits need to be delayed by placing latches to ensure arrival to the FB latches on the correct moment. To insert *C*, the signal *en_MADD* is permanently high, while *en_FB* is always low.

When going to the 40 nm technology node, a change in the topology of the FB latch was necessary: inverters were added at the MAC inputs, as can be seen in Fig. 5.11. Although various measures were taken to cope with the increased variations in the timing block (which will be addressed in the next section), in a few rare cases of intra-die simulations a timing error still occurred. Figure 5.13 shows the detailed configuration of the FB latch in the diagonal accumulation structure of the MAC. The timing signals change according to the row because of the non-overlapping clock signals *en_a* and *en_b*.

The situation where the problem occurred is the following for an uneven row: the latch below has just locked and its output signals are full-swing, as wanted. Then, the FB latch goes transparent and out of lock, but there is a slight $1-1$ overlap between *en_MAC_b* and *en_FB_b*, which is an invalid condition for the latch. If the regular (REG) latch is accidentally weaker than the FB latch due to mismatch and the bits saved in both are different, this $1-1$ overlap can occur long enough so that kickback takes place and the stored bit in the FB latch interferes with the stored bit in the REG latch, causing it to flip. The most convenient solution to avoid this unwanted kickback is to insert inverters between the outputs of the REG latch and the inputs of the FB latch, hence the kickback will never be able to cause a bit flip in the locked REG latch. This increases the energy consumption of the FB latch, but is necessary to reduce its variation sensitivity and to increase the total yield. Moreover, for an N-bit MAC, only N feedback latches are required. In comparison to the large amount of REG latches, this solution has only a very limited impact on the total energy consumption.

5.2.3.3 Timing

This section will first discuss the general functionality of the timing of the MAC and will then elaborate on the various technology differences in implementation.

Figure 5.14 shows the implementation of the timing used for the MAC. There are three inputs for the timing: the input clock *clock_in* from which the non-overlapping clocks are deduced, and the two previously mentioned configuration bits *reset_in* and *madd_in* to determine the operation mode (explained in Table 5.2). The outputs consist of the clock signals for the REG latches *en_a* and *en_b*, clock signals *en_MAC_a/b* and *en_FB_a/b* for the FB latches, as well as the enable signal for the MADD mode *en_MADD* and the reset signal $\overline{reset}$.

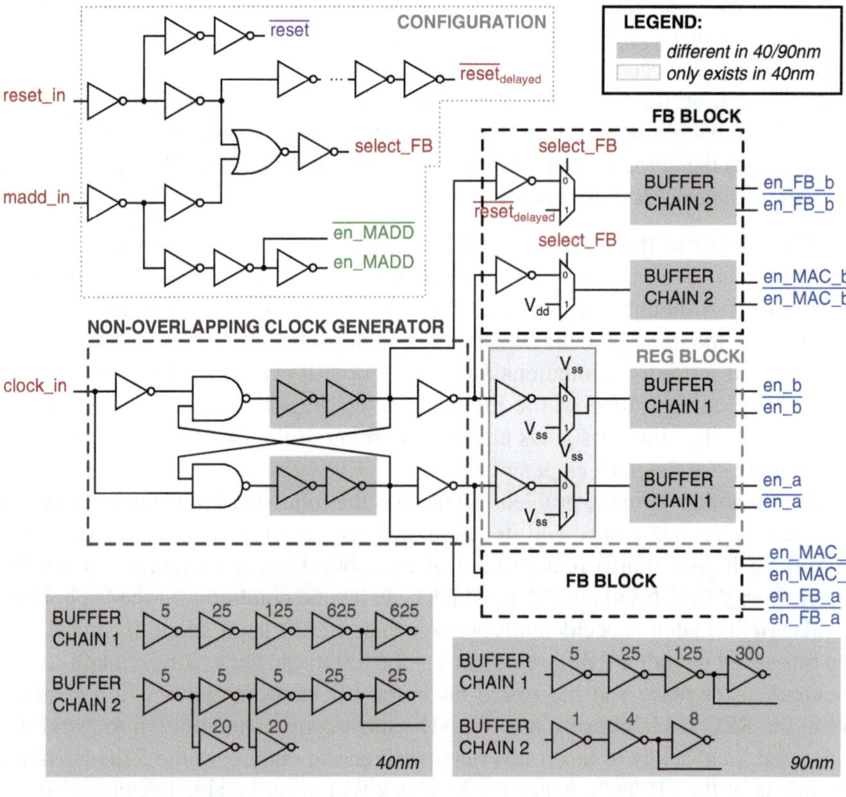

Fig. 5.14 Implementation of the timing [10]

Table 5.2 Configuration bits per operation mode [10]

Operation mode	*reset_in*	*madd_in*
MAC	0	0
MULT	1	0
MADD	0	1

The internal signals *select_FB* and $\overline{reset}_{delayed}$ are used to configure the timing signals of the FB latch. The amount of delay that is inserted for $\overline{reset}_{delayed}$ is determined by process corner simulations. As explained before, this inserted delay needs to ensure that the signal levels of the reset nodes of the FB latch are settled before establishing the cross-coupled connection. Therefore, the delay of the inverter chain is determined to be higher than the maximal rise and fall time of these nodes in all process corners.

The timing block functions at the same supply voltage V_{dd} as the MAC. The logic gates used in the timing are implemented as standard CMOS logic gates and not as TG logic gates because differential input signals were not available and only a few logic gates were needed. More precisely, the NAND gate is implemented as a regular standard CMOS NAND with appropriate (and therefore increased) sizing of the pMOS, and the NOR gate is implemented using stacked nMOS transistors to reduce the required pMOS sizing.

Some of the blocks of the timing are implemented differently in both technologies. This comes from the significant increase in variations in the 40 nm technology, which introduced many extra challenges. The 90 nm technology was significantly less sensitive to variations. Note that all inverters of which the sizing is not explicitly mentioned in Fig. 5.14 were implemented minimally (a relative sizing of 1) in the 90 nm node, whereas in the 40 nm case they were implemented with a relative sizing of 5 to reduce the sensitivity to variations. This upsizing is only used in the timing block, the MAC implementation is sized as explained earlier.

The main considerations that had to be taken into account when designing the timing were:

- **Ensure the non-overlap time between the REG clock signals:** The non-overlap time between *en_a* and *en_b* is controlled by the non-overlapping clock generator. The gray shaded area herein indicates the inverter chain which is inserted to increase the non-overlap time of the clock signals. So that under all variations there would never occur any overlap, a chain of 4 respectively 6 inverters was sufficient for 90 nm and 40 nm.
- **Ensure the non-overlap time between the FB clock signals:** In MAC mode, the FB latches work as REG latches and therefore a non-overlap time between *en_MAC* and *en_FB* is required. This is ensured by matching the paths of both signals as good as possible, which was obtained by having the exact same amount of logic gates and inverters in both paths. To reduce mismatch, the muxes in the FB block are implemented as TG muxes. This was possible because only the controlling signal of a TG mux needs to be differential.

- **Ensure the non-overlap time between the REG and the FB clock signals:** Since the FB latches work as regular latches in MAC mode, it is imperative that there is no overlap between *en_a/b* and *en_MAC_b/a* (refer also to Fig. 5.13). In the 90 nm technology, this proved to be not an issue. In the 40 nm node however, matching of the paths was crucial to satisfy this requirement. A few measures were taken to match the paths of the regular and the FB clock signals as meticulous as possible. First of all, dummy inverters and muxes were inserted in the regular path (Reg Block in Fig. 5.14) to match the delay of the same elements in the FB Block. Additionally, buffer chain (BC) 2 of the FB Block needed to be matched to BC 1 of Reg Block. Simply increasing the sizing of the buffers would unnecessarily increase the energy consumption. Therefore, BC 2 was matched by adding buffers that increased the fan-out of the previous buffer but that were not present in the signal path, thus allowing to not increase the size of the following buffer.

5.2.4 Measurement Results

Figure 5.15 shows the micrographs of both chips. The active area of the 90 nm version is $310 \times 225\,\mu m^2$, while the 40 nm one is $153 \times 153\,\mu m^2$. This corresponds to an area reduction of 66 %. According to the classical scaling law, a reduction of around 80 % is expected, but this law does not apply anymore since the transistor area and wiring pitch do not scale likewise in advanced nanometer CMOS

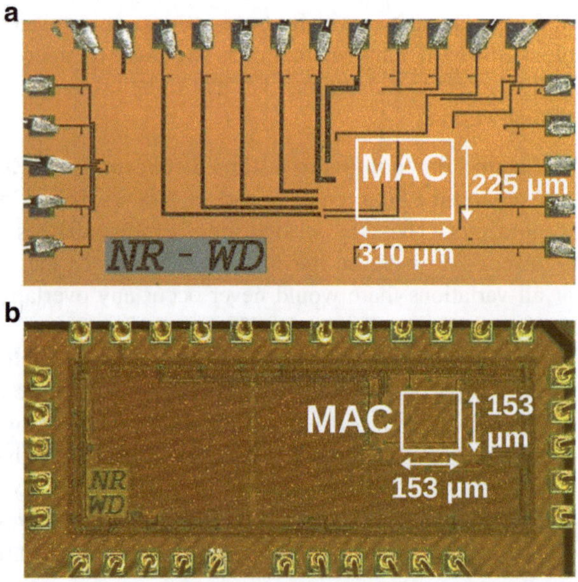

Fig. 5.15 Chip micrograph of the 16-bit MACs: (**a**) 90 nm MAC [9] and (**b**) 40 nm MAC [10]

Table 5.3 Comparison of measurement results of the 16-bit MAC in both technologies [10]

CMOS Technology		90 nm	40 nm	Difference
# of measured dies		34	20	
Active area	[μm^2]	225 × 310	153 × 153	−66 %
$V_{dd,min}$	[mV]	150	180	+20 %
Clock frequency	[MHz]			
@ $V_{dd,min}$		5.0	12.0	
@ 190 mV		10.48	17.06	+63 %
@ 250 mV		31.88	53.48	+68 %
σ/μ @ 190–290 mV		16.77 %	11.93 %	−29 %
Energy/operation	[pJ]			
@ $V_{dd,min}$		0.97	1.43	
@ 190 mV	(MEP)	0.87	1.32	+51 %
@ 250 mV		1.10	1.61	+46 %
σ/μ @ 190–290 mV		7.94 %	6.10 %	−23 %
EDP	[pJ. μs]			
@ 190 mV		0.088	0.079	−11 %
@ 250 mV		0.035	0.031	−14 %
Leakage power	[μW]			
@ 190 mV		3.90	14.18	+364 %
@ 250 mV		5.96	26.60	+447 %

technologies. Scaling according to transistor area results in a reduction of 50 % and according to wiring pitch −44 %. To conclude, the 40 nm version has scaled exceptionally well compared to the 90 nm version.

All measurement results provided below are for the MAC mode. Measurements of the two other modes produce very similar results. To study the variation-resilience of both designs, a significant number of dies was measured in both technologies: 34 dies for the 90 nm case and 20 for the 40 nm case. All important specifications of the measurement results are summarized in Table 5.3.

Figure 5.16 provides the distribution of the measured minimal supply values of both chips at which the dies were still functional. Measurement results show that $V_{dd,min}$ of the 90 nm MAC is 150 mV, and the 40 nm MAC is able to work down to 180 mV. The measurements thus demonstrate a very good match with the practical $V_{dd,min}$ values of 158 mV and 186 mV respectively, which were theoretically derived in Sect. 2.3.1.

As derived from Fig. 5.16, the mean values of $V_{dd,min}$ are 182 mV for the 90 nm MAC ($\sigma = 19.7$ mV) and 200 mV for the 40 nm version ($\sigma = 12.1$ mV).

Figure 5.17 compares measurement results of the first and the second prototype of this book. Since these were fabricated in the same technology, a frequency comparison between the two designs provides meaningful insights in the impact of the different optimized design decisions for the MAC design. Remember that the main optimizations are the differential implementation of TG logic, the fully

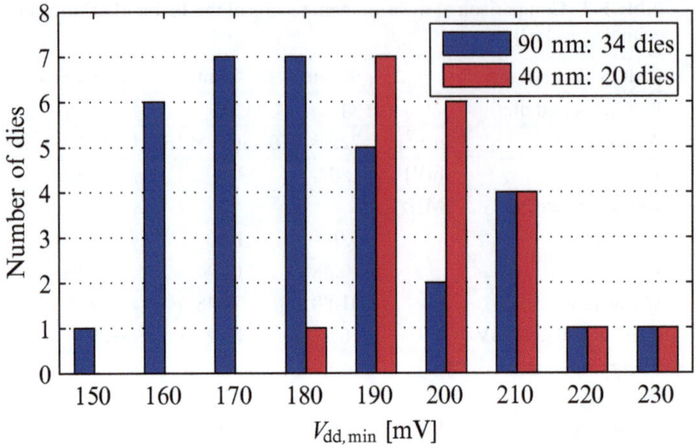

Fig. 5.16 Distribution of the minimal functional supply voltage $V_{dd,min}$ of the measured dies in both technologies

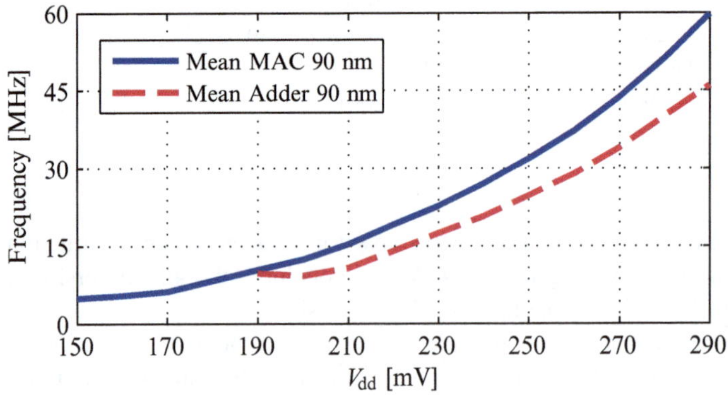

Fig. 5.17 Comparison of mean results of clock frequency measurements of multiple dies of the first prototype (90 nm adder) and the second prototype (90 nm MAC), as function of V_{dd}

differential latch and the increased pipeline stage length from 1 to 2 logic gates, which are all expected to drastically increase variation-resilience. First of all, this is verified by the fact that the MAC is functional down to a significantly lower $V_{dd,min}$ of 150 mV, compared to the 190 mV of the adder, which can be attributed to the enhanced variation-resilience. Second, the frequency comparison of Fig. 5.17 shows that for the same supply voltage, the MAC is able to operate at a clock frequency equal or higher than the one of the adder. The impact of timing variations is thus drastically reduced, because the clock frequency improves at the same supply voltage, although the pipeline stage length is doubled in the MAC.

The remainder of this section will focus on the comparison of the measurement results of both MACs in different technologies. The upper plot of Fig. 5.18 shows the

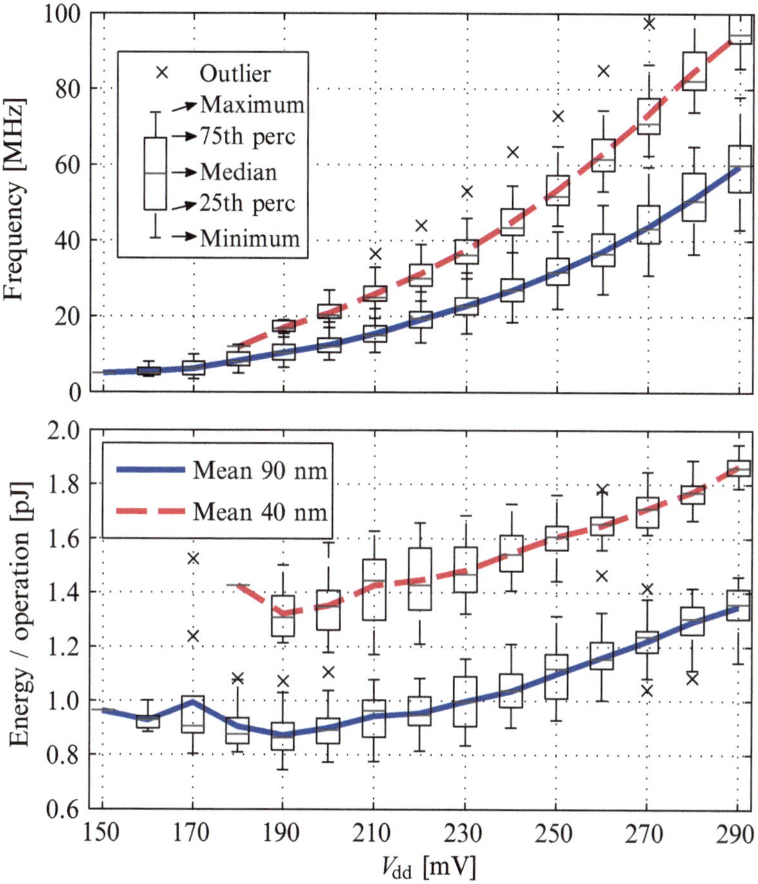

Fig. 5.18 Boxplot of the measured maximum clock frequency and energy consumption per operation as function of V_{dd}

measured maximum operating frequencies at which each die was able to function correctly at a certain V_{dd}. At 190 mV, a mean clock frequency of 17.06 MHz is obtained with the 40 nm MAC, which is 63 % higher than the mean frequency of 10.48 MHz of the 90 nm MAC at that supply. For a supply ranging from 190 to 290 mV, the 40 nm MAC achieves a mean clock frequency improvement of approximately 66 % over the 90 nm MAC. In the same supply range, the mean variation σ/μ is 11.93 % for 40 nm and 16.77 % for 90 nm, thereby illustrating the variation-resilience of both designs. To conclude, the 40 nm MAC is able to operate significantly faster than the 90 nm MAC for the same supply voltages. Moreover, it also achieves a better variation-resilience than the 90 nm MAC, with a reduction of 29 %. This can be explained by the fact that for the same V_{dd}, the 40 nm transistors are further away from the sensitive sub-threshold region due to their lower V_T. Finally, these measurement results thus show that technology scaling is beneficial for sub- and near-threshold circuits in terms of clock frequency. Note also that the expected decrease of delay is accomplished when going from 90 nm to the smaller 40 nm technology, as opposed to what the simulations suggested (recall Sect. 2.3.3).

The total energy consumption per MAC operation is shown in the lower plot of Fig. 5.18. Extra on-chip circuitry makes it possible to do at-speed energy consumption measurements while applying arbitrary inputs (recall Sect. 4.5). The MEP of both designs coincides at 190 mV, where the 40 nm MAC consumes 1.32 pJ per operation, which is an increase of 51 % compared to the 0.87 pJ of the 90 nm MAC. For the 190 to 290 mV supply range, the energy consumption of the 40 nm version is 46 % higher than the 90 nm MAC.

Whereas the variation-resilience still improves, unfortunately the energy consumption deteriorates considerably. This total energy consumption consists of a static and a dynamic component. The dynamic energy is expected to decrease with scaling, and the static or leakage energy will increase. In this comparison, the dynamic energy scaling does not completely follow the ideal scaling laws, as the sizing of the basic building blocks (see Sect. 3.4) and the timing (see Sect. 5.2.3.3) is changed. Moreover, in reality, technologies do not scale according to the ideal scaling laws e.g. wire capacitances do not scale as well as transistor capacitances. Regarding the leakage component, Fig. 5.19 provides the measured absolute leakage power and relative contribution to the total power consumption as function of V_{dd}. The increased total energy consumption can mainly be attributed to the increased leakage. At a supply of 190 mV, the leakage contribution of the 90 nm MAC to the total power consumption is 43 %, while the 40 nm MAC is dominated by a leakage contribution of 63 % at that point. Moreover, not only the relative leakage contribution increases, but the absolute leakage power increases drastically as well, e.g. from 3.9 μW to 14.2 μW at 190 mV. The leakage component in the total power consumption thus increases substantially for the same supply voltage, thereby explaining the increased total energy consumption.

The question is now: is it advisable to go to advanced nanometer technologies for ultra-low-voltage designs? From an area perspective, it certainly is. Furthermore, the operating frequency increases drastically, but so does the energy consumption per operation. To be able to make a fair comparison between these last two metrics, the EDP is the adequate FOM. The EDP as function of the supply voltage is given in Fig. 5.20. The EDP is calculated as the energy consumption per operation divided by the clock frequency (or throughput) of the MAC. It is not calculated by multiplying energy by the total latency, since the throughput is the metric that indicates the number of inputs that can be calculated in a certain time period for pipelined systems. In terms of EDP, the 40 nm design outperforms the 90 nm version, with a reduction of 13 % for the 190 to 290 mV supply range.

To conclude, in an application where energy consumption is of vital importance and speed is of much lower concern, the 90 nm version is the more suitable technology of both, whereas from an area point of view, the 40 nm version is recommended. From an EDP perspective, it depends on whether the ultra-low-voltage design is used in an application or a larger system with a fixed supply voltage, where the 40 nm version performs better at a single given supply voltage, or whether the

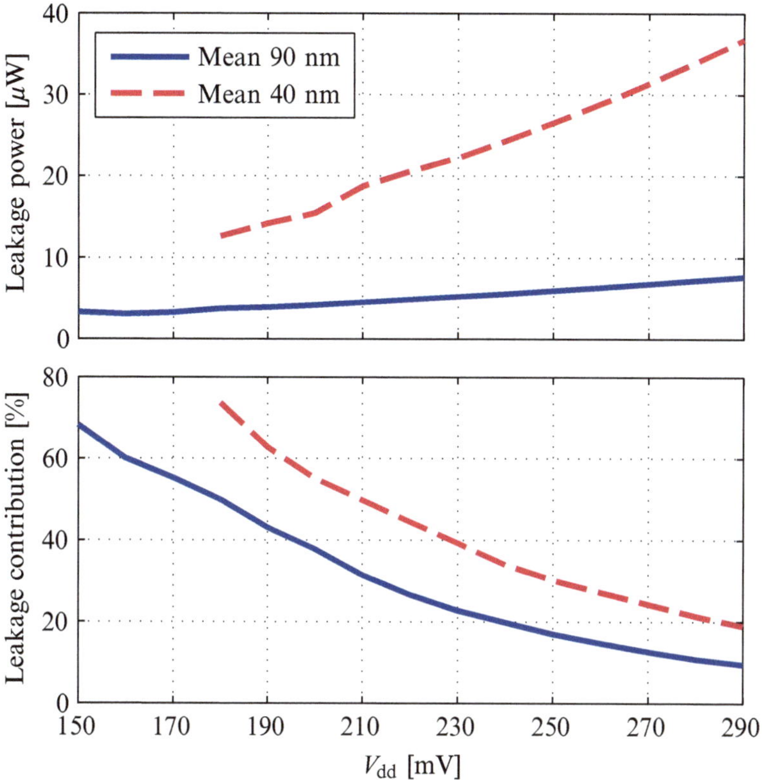

Fig. 5.19 Measured leakage power and contribution of the leakage to the total power consumption as function of V_{dd}

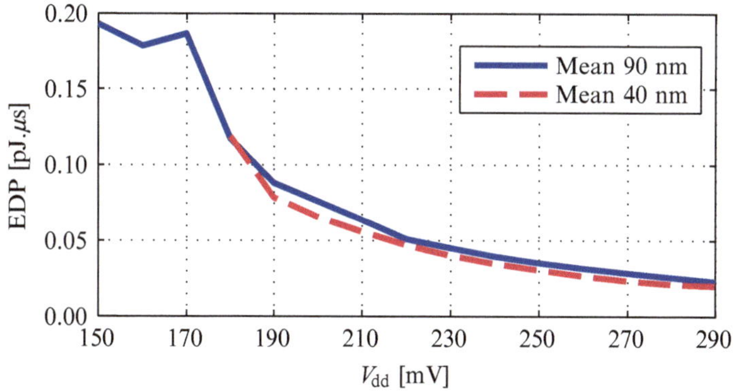

Fig. 5.20 Measured EDP as function of V_{dd}

supply voltage can be freely chosen. In the latter case, it is possible to operate the 90 nm MAC at the same frequency and a slightly higher supply voltage as the 40 nm version, while achieving a lower energy consumption and hence a lower EDP.

Previous work also predicted that the static energy increases dramatically with technology scaling and that this trend can compromise scaling benefits. However, the prediction of the technology node at which this compromising point will occur differs. As stated in [2], the minimum energy consumption increases for the same design when going from a 90 nm technology to a 45 nm node. In [1], it is found that the static energy increase will specifically be dramatic at the 32 nm node and that the benefit of scaling in terms of energy consumption will start to diminish for 45/32 nm technology nodes and below. The authors of [11] stated that by scaling technology from 0.25 μm to 65 nm, the energy consumption can be reduced significantly, but that from an energy consumption point of view, there is no clear benefit to use technologies smaller than 45/65 nm for ultra-low-power purposes. The authors predict that the EDP will start to slowly increase at the 32 nm node.

With the measurement results of the MAC designs, it is possible to conclude that, in terms of the energy-performance trade-off, ultra-low-voltage circuits should be scaled down to advanced nanometer technologies, at least until the 40 nm node. The benefits of further scaling depend on both the increased leakage, as well as the increased variations. The domination of static leakage of ultra-low-voltage designs in advanced nanometer technologies has consequences for the future of scaling. If, for future technologies below 40 nm, the leakage becomes too high compared to the increase in speed, there will not be any improvement in EDP at a given supply anymore for ultra-low-voltage circuits. More precisely, there will come a point in scaling when the gain of the decreased dynamic energy consumption will be outweighed by the increase in static leakage and the increased variability. If the advance in speed is not able to compensate this, the EDP will not reduce and there will be nothing to gain from further scaling. In this 40 nm technology, the EDP improves because the balance between speed gain, dynamic energy reduction and static leakage increase is still positive.

5.2.5 State-of-the-Art Comparison

Table 5.4 provides a state-of-the-art-comparison of ultra-low-voltage MACs. Making a comparison between this work and the only other published ultra-low-voltage MAC [6] is difficult because it is not processed in the same CMOS technology and it has a different bit length. As shown in the technical specs, the MAC in [6] has been produced in a 0.14 μm CMOS technology. Therefore, the comparison is made with the 90 nm version of the MAC since that is the one closest to the technology of the referenced paper. The architecture differs as well: the referenced design consists of an 8-bit sequential MAC, i.e. an 8-bit array multiplier followed by a standard 24-bit ripple carry adder and accumulator.

Table 5.4 State-of-the-art ultra-low-voltage MAC comparison [9]

		This work	[6]
Technical specs:			
Bit length MAC		16-bit	8-bit
Technology node		90 nm	0.14μm
Architecture		Pipelined, Interwoven	8-bit Multiplier + 24-bit Adder
Measured results:			
$V_{dd,min}$	[mV]	150	175
Throughput	[MHz]	5.0	0.166
Power	[μW]	4.8	0.014
Energy	[pJ]	0.96	0.084
EDP	[pJ.μs]	0.193	0.508

Table 5.5 Transferred state-of-the-art ultra-low-voltage MAC comparison, to acquire equal bit length and technology, for operation at $V_{dd,min}$ [9]

Specifications		This work			[6]			
@ $V_{dd,min}$		16-bit	→	8-bit	140 nm	→	90 nm	Difference
V_{dd}	[mV]	150	=	150	175	=	175	+17 %
Throughput	[MHz]	5	=	5	0.166	$\times S$	0.258	−95 %
Energy	[pJ]	0.96	÷4	0.24	0.084	$\div S$	0.054	−78 %
EDP	[pJ.μs]	0.193	÷4	0.048	0.508	$\div S^2$	0.210	+438 %

$S = 140 \div 90 = 1.556$

A comparison has been made on reported specifications at the minimal supply voltages of both designs, shown in the measured results of Table 5.4. [6] is able to operate at a minimal V_{dd} of 175 mV, while this work is functional down to 150 mV. In order to get a meaningful comparison, the specifications have been reworked in Table 5.5 to acquire equal bit length and technology. First, the measurement results of this work have been transferred from a 16-bit to an 8-bit design. Second, the results of [6] are ported from a 140 nm to a 90 nm CMOS technology in an optimistic manner, without taking into account the increased leakage. This comparison shows that while the referenced MAC reaches a smaller energy consumption, it also has a significantly smaller operating frequency. Therefore, this work has a 4.38 times better EDP for $V_{dd,min}$. To conclude, this design outperforms [6] in performance and EDP.

5.2.6 Conclusion

This section described the design of a 16-bit MAC. The targets of the design of the second and third prototype were twofold. First, improvements on both the architectural design and the gate-level building blocks have been implemented in the design of the MAC to enhance the overall variation-resilience. These changes—differential TG logic, fully differential latches and increased pipeline stage length—resulted in an improved clock frequency, reduced minimal supply voltage and increased variation-resilience, and were therefore successfully validated.

Second, the MAC has been used as a test vehicle to study the effects of CMOS technology scaling. The design changes which were necessary for the technology scaling have been extensively discussed. Afterwards, the measurement results have been compared in detail in order to determine the benefits and disadvantages of scaling of ultra-low-voltage circuits. The measurements demonstrate a drastically improved operating frequency at the cost of a higher energy consumption, resulting in a reduced EDP at a given supply voltage for the 40 nm MAC. Scaling to the 40 nm node is beneficial for ultra-low-voltage designs in terms of area, in terms of operating frequency and EDP at a fixed V_{dd}, but not in terms of energy consumption. It is shown that the effect of scaling on the EDP for such designs will be positive as long as the static leakage is kept under control. Although advanced nanometer CMOS technologies suffer from an increased variability, measurements show that both MACs are still variation-resilient.

5.3 Conclusion

The design strategy and methodology which has been thoroughly discussed in the previous chapters has been implemented in three different prototypes in this chapter. Successful measurements of these prototypes have validated the proposed gate-level building blocks and architecture. Extensive comparison between the results of the three chips allowed to not only implement significant improvements, but also to research the influence of CMOS technology scaling. The obtained insights will be used for the design of the fourth and final prototype of this book in the following chapter.

References

1. Bol D, Ambroise R, Flandre D, Legat JD (2009) Interests and limitations of technology scaling for subthreshold logic. IEEE Trans Very Large Scale Integ (VLSI) Syst 17(10):1508–1519. DOI:10.1109/TVLSI.2008.2005413
2. Bol D, Kamel D, Flandre D, Legat JD (2009) Nanometer MOSFET effects on the minimum-energy point of 45nm subthreshold logic. In: Proceedings of the ACM/IEEE international symposium on low power electronics and design (ISLPED), pp 3–8. DOI:10.1145/1594233.1594237
3. Brent RP, Kung HT (1982) A regular layout for parallel adders. IEEE Trans Comput C-31(3):260–264. DOI:10.1109/TC.1982.1675982
4. Calhoun B, Chandrakasan A (2006) Ultra-dynamic voltage scaling (UDVS) using sub-threshold operation and local voltage dithering. IEEE J Solid State Circuits 41(1):238–245. DOI:10.1109/JSSC.2005.859886

5. Hatamian M, Cash G (1986) A 70-MHz 8-bitx8-bit parallel pipelined multiplier in 2.5-μm CMOS. IEEE J Solid State Circuits 21(4):505–513. DOI:10.1109/JSSC.1986.1052564
6. Kao J, Miyazaki M, Chandrakasan A (2002) A 175-mV multiply-accumulate unit using an adaptive supply voltage and body bias architecture. IEEE J Solid State Circuits 37(11):1545–1554. DOI:10.1109/JSSC.2002.803957
7. Kogge PM, Stone HS (1973) A parallel algorithm for the efficient solution of a general class of recurrence equations. IEEE Trans Comput C-22(8):786–793, DOI:10.1109/TC.1973.5009159
8. Reynders N, Dehaene W (2011) A 190mV supply, 10MHz, 90nm CMOS, pipelined sub-threshold adder using variation-resilient circuit techniques. In: Proceedings of the IEEE Asian solid-state circuits conference (A-SSCC), pp 113–116. DOI:10.1109/ASSCC.2011.6123617
9. Reynders N, Dehaene W (2012) Variation-resilient sub-threshold circuit solutions for ultra-low-power digital signal processors with 10MHz clock frequency. In: Proceedings of the IEEE European solid-state circuits conference (ESSCIRC), pp 474–477. DOI:10.1109/ESSCIRC.2012.6341358
10. Reynders N, Dehaene W (2015) On the effect of technology scaling on variation-resilient sub-threshold circuits. Elsevier Solid State Electron 103:19–29
11. Tajalli A, Leblebici Y (2011) Design trade-offs in ultra-low-power digital nanoscale CMOS. IEEE Trans Circuits Syst Regul Pap 58(9):2189–2200. DOI:10.1109/TCSI.2011.2112595

Chapter 6
JPEG Encoder

This chapter presents the design and measurement results of the fourth and final ultra-low-voltage prototype. Since signal processing applications are the focus of this book, a JPEG encoder is chosen as a representative DSP block to validate the design strategy which has been proposed in all the previous chapters. The purpose of the JPEG encoder is to demonstrate that the proposed circuit and architectural techniques are generally applicable in any large and complex ultra-low-voltage Digital Signal Processor (DSP) design. The targets of this prototype remain unchanged: operation at ultra-low supply voltages to enable a high energy-efficiency, operating frequencies in the range of $n \times 10\,\text{MHz}$ and high variation-resilience. The design efforts which are made to accomplish these targets are profoundly discussed in this chapter. The JPEG encoder is fabricated in a 40 nm CMOS technology [17].

Section 6.1 explains why the JPEG encoder is chosen as a proof of concept for any large DSP block, while Sect. 6.2 provides an overview of the JPEG encoding algorithm and the division of the various subblocks which realize this algorithm. The general architectural and gate-level design choices for this ultra-low-voltage prototype will be discussed in Sect. 6.3.

Section 6.4 covers the detailed design of the subblocks of the JPEG encoder, with a focus on how the research targets—high energy-efficiency, speed and variation-resilience—are achieved. The measurement results will be examined profoundly in Sect. 6.5, whereas Sect. 6.6 provides an extensive state-of-the-art comparison between all published ultra-low-voltage processors in advanced nanometer CMOS technologies.

Provided that some initial design targets of this prototype are revisited, improvements are possible in the lookup tables of this JPEG encoder. These will be discussed in Sect. 6.7, where these tables are used as a case study to explore circuit techniques for increasing their energy-efficiency [18]. Finally, Sect. 6.8 concludes this chapter.

N. Reynders, W. Dehaene, *Ultra-Low-Voltage Design of Energy-Efficient Digital Circuits*, Analog Circuits and Signal Processing,
DOI 10.1007/978-3-319-16136-5_6

6.1 Proof of Concept

The aim of this chapter is to design a complete, sufficiently large DSP block that advances the state-of-the-art by not only reaching very low energy consumption, but also clock frequencies of tens of MHz, while providing high variation-resilience. Since a JPEG encoder is a representative DSP block, it serves as an interesting design case to validate the proposed design strategy of this book for any DSP application. The JPEG encoder consists of a large datapath, control logic and requires memory. It is fabricated in a 40 nm CMOS technology [17].

The design of this fourth prototype builds further upon the insights acquired when designing and measuring the three datapath prototypes of Chap. 5. Moreover, this prototype reuses the adder and the MAC design, as will be discussed later. Furthermore, the design strategy is expanded to control logic and memory as well. A focus is given on how to implement this control logic and memory in a highly energy-efficient manner for ultra-low-voltage operation. The intention of this prototype has been to operate at a slightly higher supply than the previous prototypes, in order to be able to have a longer pipeline stage length while guaranteeing robustness.

6.2 JPEG Encoding Algorithm

Figure 6.1 visualizes the functionality of the JPEG encoder, as well as its main building blocks. The implemented JPEG encoder is compliant with the baseline sequential mode of the Joint Photographic Experts Group (JPEG) image compression standard [20]. In the JPEG algorithm, an image is split into blocks of 8 by 8 pixels. These pixels are represented by 8-bit integers.

The 8×8 blocks are then used as inputs for the first building block, which is the two-dimensional Discrete Cosine Transform (DCT). This 2D-DCT transforms the blocks of 8×8 pixels to the frequency domain. The upper left coefficient of the DCT output is the DC frequency, while the 63 other coefficients are AC coefficients. The AC frequencies in the upper left corner are the lower frequency values, while the 63rd coefficient in the lower right corner is the highest frequency component. The advantage of performing DCT is that most of the energy of the original block is concentrated in the lower frequency components of the DCT output. Therefore, the higher frequency components have very low values, equal to or slightly higher than zero. As can be seen in Fig. 6.1, the 2D-DCT temporarily increases the number of bits as the DCT coefficients are represented by 15-bit numbers.

The following building block is the so-called quantization, which defines the amount of compression of the JPEG image. In the quantization, the block of DCT coefficients is divided by a quantization matrix of 8×8 coefficients, which is stored in the quantization table. Each DCT coefficient is scaled with a separate quantization coefficient, because the sensitivity of the human eye differs for different frequencies,

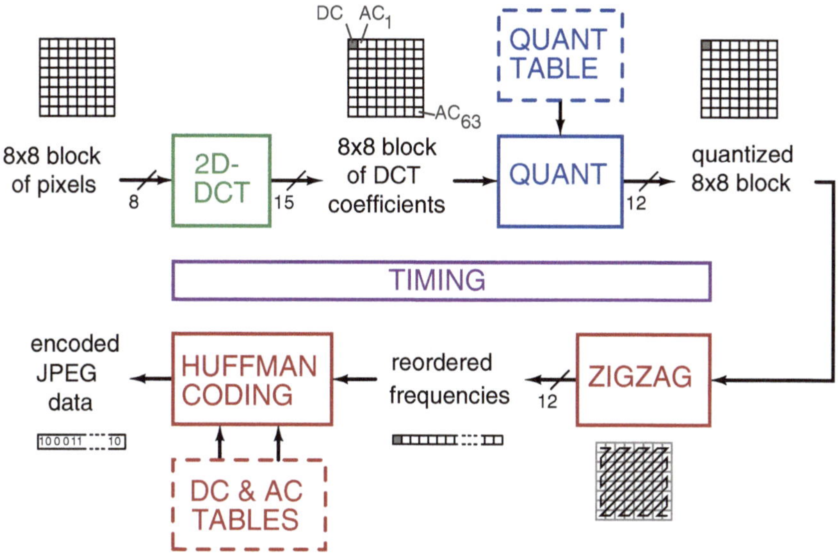

Fig. 6.1 Block diagram of the JPEG encoder

i.e. it is much more sensitive to low frequency components than it is to higher frequency components. Larger values of quantization coefficients provide greater compression. To summarize, this building block controls the compression rate and thus the quality of the final image. After the quantization, the number of bits needed to represent all possible values has reduced to 12-bit.

In the following step, the quantized 8 × 8 block is reordered. This happens in a zigzag-order so as to group similar frequencies. This array of frequencies can then be efficiently coded by the Huffman encoding. After the quantization, the amount of zero values has increased considerably, especially in the higher frequency components. Huffman encoding takes advantage of this by only coding non-zero AC values and by simultaneously including the amount of preceding consecutive zeros, which is called *runlength*, within this code. Huffman codes do not have a predefined length, but rather variable lengths. These lengths are based upon the estimated occurrence of the combination of a non-zero value and a certain runlength of zeros. DC and AC frequencies are encoded separately, according to the codes provided in the DC and AC Huffman tables, respectively.

To conclude, essential to know is whether a building block contains simply a logic function or if it also needs memory. The 2D-DCT can be seen as a large and complex datapath, while both the quantization and the zigzag & Huffman coding blocks have relatively less logic but require stored data in lookup tables.

6.3 Ultra-Low-Voltage Design

From an *architectural* point of view, the JPEG encoder has a latch-based pipelined architecture, as do the three other prototypes. Deep pipelining is used in this prototype as well, although the pipeline stage length has been increased in comparison with the MAC design in the same 40 nm CMOS technology. Since the 40 nm MAC has been employed to study technology scaling, the pipeline stage length had to remain equal to the 90 nm Multiply-Accumulate Unit (MAC) to allow a fair comparison. Hence, the same pipeline stage length of 2 logic gates as the 90 nm MAC was used.

The aim of the JPEG encoder was to increase the pipeline stage length in order to cope with timing variations even more effectively. Performing the analysis of Sect. 4.2.2 for the 40 nm technology at hand revealed that the maximum allowable logic depth could be increased to three logic gates provided that the target supply voltage also increased slightly. According to the simulation results of the analysis, the target $V_{\text{dd,min}}$ should be 230 mV. Therefore, this value has been used for the functional verification simulations of the system and its subblocks. The measured $V_{\text{dd,min}}$ will turn out to be 210 mV, as will be shown later in the measurement results of Sect. 6.5. To conclude, a maximal pipeline stage length of three consecutive logic gates has been used for the design of the JPEG encoder.

From a *gate-level* perspective, the implementation choices of logic gates and latches are the same as for the 40 nm MAC, as those were successfully validated during measurements. Differential TG logic has been employed, and the sizing of the logic gates and the latches is as discussed in Sect. 3.4. The latch has again been implemented in a fully differential manner.

6.4 Implementation

In this section, the different building blocks will be discussed in detail. Note that this entire JPEG encoder has been designed in a custom design flow with transistor-level simulations.

6.4.1 Timing

The JPEG encoder consists of a latch-based pipelined architecture. The entire chip functions in a single voltage and clock domain. The clock is distributed throughout the chip by a clock tree. Non-overlapping clocks (*en_a* and *en_b*) are used to avoid possible race conditions. The timing block in Fig. 6.1 thus consists of a Non-Overlapping Clock Generator (NOCG) and the complete clock tree, as visualized in Fig. 6.2.

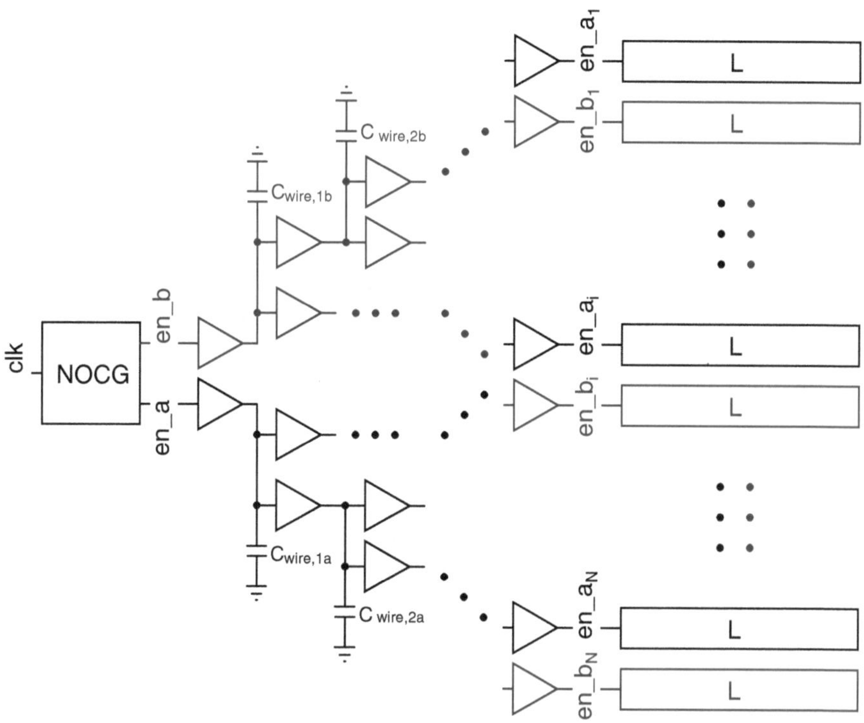

Fig. 6.2 Implementation of the timing of the JPEG encoder

The implementation of the NOCG has been discussed in Sect. 4.3.3. It is equal to the one used in the 40 nm version of the MAC (as explained in the timing part of Sect. 5.2.3), as this is the same technology and the measurements of that chip verified successful functionality.

The clock tree is designed in a fully custom manner, using parasitic extractions to determine wire capacitances. The NOCG is situated left to the JPEG encoder in the middle of the layout. From there, the enable signals are distributed by the clock tree to the rows of latches. Recall Fig. 4.13 which shows the systematical layout structure employed for the prototypes: each latch in one row is clocked by the same enable signal. Clock skew between latches of the same row is minimized because enable signals of one row are delivered by the same buffer, as shown in Fig. 6.2.

The wire capacitance of a row has been combined with the capacitance of the latches on that row to determine the total capacitance. Since the capacitance of the latches is directly proportional to the amount of latches, the row with the highest amount of latches has been used as reference case. A unit buffer of size 64 proved to be sufficient to drive that worst-case row with a fan-out of 4. This unit buffer has then been used in the entire clock tree. An optimization has been carried out for rows which contained only a small amount of latches: the buffer is shared between two rows in that case. Using this methodology, a clock tree of depth 6 has been obtained. Further timing verification has been performed with transistor-level simulations.

6.4.2 2D-DCT

Figure 6.3 shows the implementation of the 2D-DCT. The separability property of the 2D-DCT allows the transform to be calculated one dimension at a time. Therefore, the 2D-DCT is implemented as a sequence of two 1-dimensional DCTs with a transpose matrix in between: the first 1D-DCT is calculated row-wise, and the second in column-order.

The 1D-DCT algorithm is visualized in Fig. 6.4 and is based on the algorithm proposed in [11]. It contains six calculation stages: 5 stages use addition or subtraction and 1 stage requires multiplication. The 2D-DCT is deeply pipelined, as explained before.

The implementation of the 1D-DCT is shown in Fig. 6.3. Each stage consists of a control block, a register block, operand muxes and a 15-bit adder/subtracter or multiplier. The register block has two columns: in the first column, the next eight coefficients are serially clocked in, while the second column holds the current eight coefficients for eight clock cycles to start the calculations required for the eight coefficients in the subsequent stage.

The operand muxes fetch the correct operands for the calculation at hand, and are controlled by the *sel_mux* signal from the Finite State Machine (FSM) in the

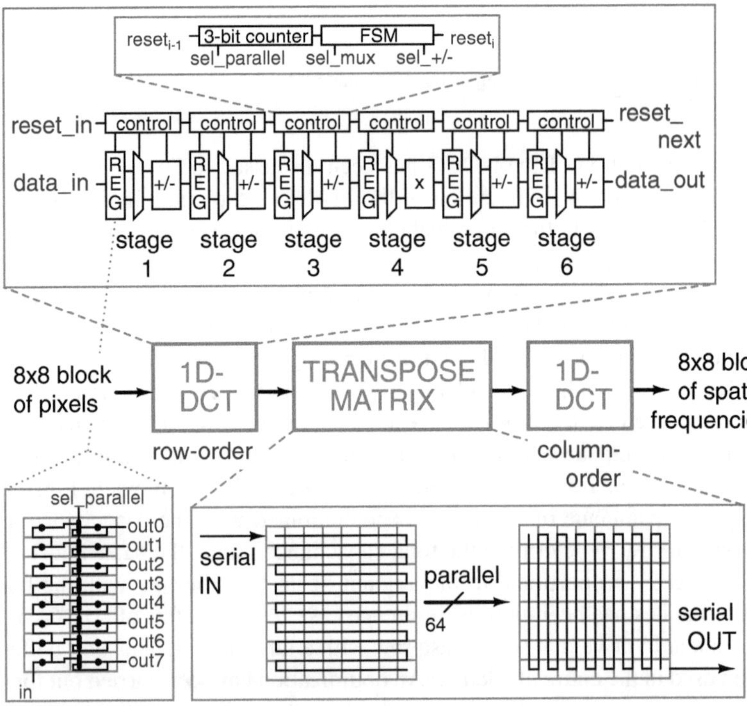

Fig. 6.3 Implementation of the 2D-DCT

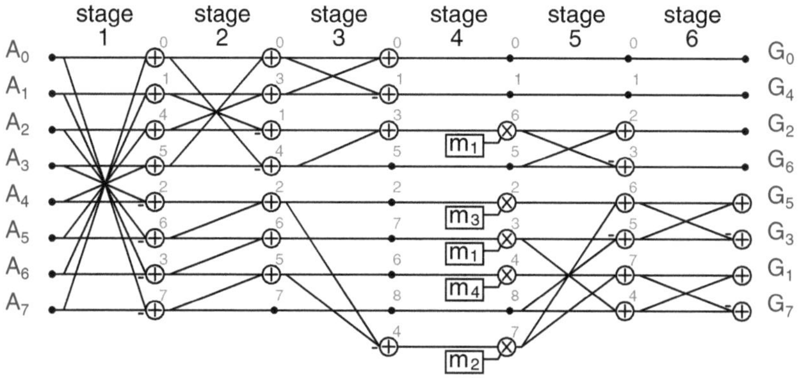

Fig. 6.4 One-dimensional DCT algorithm

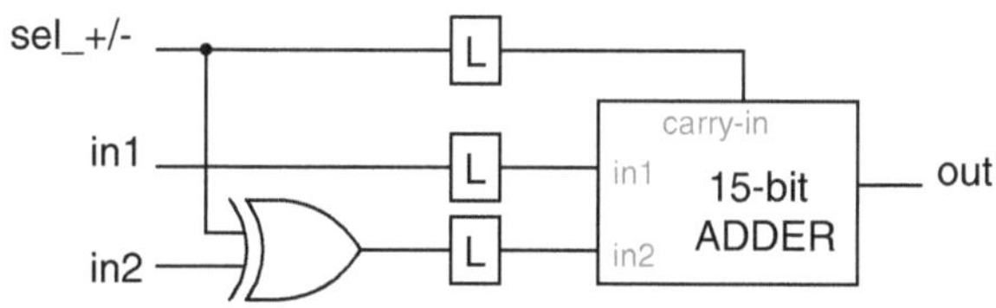

Fig. 6.5 Implementation of adder/subtracter in the 2D-DCT

control logic. The FSM also indicates through the *sel_+/-* signal whether an addition (*sel_+/-* low) or a subtraction (*sel_+/-* high) is needed, when required. The 3-bit counter keeps track of the calculation sequence and controls when to perform the parallel shift from the first to the second column in the register block, as indicated by the *sel_parallel* signal. The FSM uses the counter output to determine its current state. The counters are continuously running, and are reset by an initial *reset* signal which flows through the control logic and resets all counters at the moment of arrival of the first useful data.

The 15-bit adder used in 5 of the stages is a modified version of the Han-Carlson adder of Chap. 5. It is adapted to a 15-bit version with carry-in (for the subtraction) and is implemented with differential TG logic and pipeline stages of length 3. The implementation of the adder/subtracter is shown in Fig. 6.5. As mentioned before, its operation is controlled by the *sel_+/-* signal from the FSM.

The 15-bit multiplier is a Modified Baugh-Wooley multiplier, based on the MAC of Chap. 5, with similar implementation changes as the adder. As opposed to the addition stages where the operands consist of two coefficients which change every cycle, a certain coefficient is always multiplied with one of the fixed multiplier factors $m_{1:4}$ in stage 4. These multiplier factors come from a one-hot decoded lookup table.

In all stages but stage 1, the 1D-DCT algorithm of Fig. 6.4 consists of less than 8 calculations, since sometimes a subsequent coefficient is just a copy of a current coefficient. An option would be to power gate the calculation unit at that moment which would cost extra control logic and power gating transistors.

A bypass path to transfer the coefficient to the next stage should then be inserted. However, all this extra circuitry adds to leakage. It proved to be more energy-efficient to use the already present adder/subtracter or multiplier without inserting extra circuitry and complexity. The coefficient to copy is then simply added with 0 (output of the operand muxes is 0 when no coefficient is selected) or multiplied with 1 (implemented as an extra multiplier factor m_0).

The transpose matrix (see Fig. 6.3) consists of two blocks of 8 × 8 registers combined with a counter which provides a pulse when parallel shifting is needed. The input data is serially read in row-order in the first block, then copied in parallel to the second block from which it is serially read out in column-order. All individual registers (abbreviated to *R*) mentioned in this design consist of two consecutive latches (abbreviated to *L*), which are implemented as shown in Fig. 3.31b.

6.4.3 Quantization

Figure 6.6 provides the implementation of the quantization block. The quantization is calculated by a 15-bit multiplier which is identical to the multipliers used in the 2D-DCT.

With the implemented 2D-DCT algorithm, it is necessary to scale the 2D-DCT output by a scaling matrix. This is performed without introducing extra hardware, as it is incorporated in the quantization by scaling the quantization coefficients in advance. These scaled quantization coefficients are then stored in the quantization

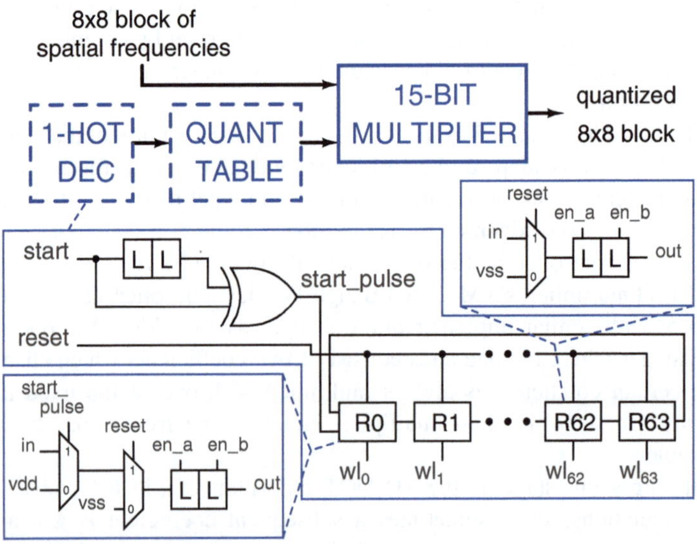

Fig. 6.6 Implementation of the quantization

table. Since all multiplier factors will be accessed in cyclic order, the table is efficiently accessed by a one-hot decoder instead of a regular decoder which would require extra input and control logic. This one-hot decoder consists of a register loop, where the first register *R0* has a different implementation than the 63 other registers to correctly synchronize the loop with the *start_pulse* signal. The one-hot decoder is reset by a *reset* signal at startup and is synchronized by a *start* signal which comes from the 2D-DCT at the moment of arrival of the first useful data.

6.4.4 Zigzag Matrix and Huffman Encoder

Figure 6.7 shows the different subblocks of the zigzag matrix and the Huffman encoder. The zigzag matrix is implemented in a similar manner as the transpose matrix, but the output of the second block is now serially shifted out in zigzag-order (as visible in Fig. 6.1). The various subblocks of the Huffman encoder will now be discussed in detail. The subblock division is loosely based on [11], but the detailed implementation is very different and many optimizations have been performed. In the following figures, the Δ symbol indicates a delay line of x latches with x given as PP $= x$.

The output of the Huffman encoder consists of two words: $\langle$*ehufco*, *amplitude*$\rangle$. The first word is the Huffman code to represent a coefficient and the second word is the amplitude of that coefficient and is only included if the coefficient is not equal to zero. Both have variable lengths, their bit lengths are therefore indicated by the Huffman size symbol *ehufsi* and by the so-called magnitude *category* symbol, respectively.

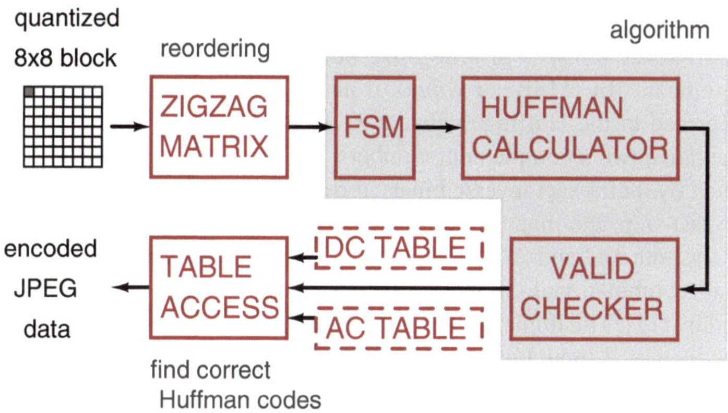

Fig. 6.7 Implementation of the zigzag block and the Huffman encoder

6.4.4.1 FSM

The FSM generates signals necessary for the Huffman algorithm, which are visible as inputs for the Huffman calculator in Fig. 6.8. The *is_dc* and *is_last_ac* signals indicate if the current value coming from the zigzag block (called *value_zz*) is coefficient 0 or coefficient 63 of the block, respectively. Both signals can be derived from the counter output of the zigzag matrix. The *is_last_block* signal is a global signal which indicates if the current block is the last 8×8 block to encode, a signal necessary for the upcoming valid checker block. Two reset signals, i.e. *reset_dc_prev_reg* for Huffman calculator and *reset_valid_checker* for valid checker, are derived from the reset signal of the zigzag counter.

6.4.4.2 Huffman Calculator

The Huffman calculator block performs calculations necessary for the encoding further on, and realizes the different steps needed for DC or AC values, as required by the JPEG standard [20]. Four different steps can be distinguished in the Huffman calculator: store previous DC value, calculate DC difference, decrement negative value and the category selection & zero runlength detector step. These will all be discussed in detail.

While AC values are encoded as such, for DC values the difference between the current DC value and the DC value of the previous block is encoded. Since there is usually a strong correlation between the DC values of adjacent blocks and DC values frequently contain a significant fraction of the total image energy, this special treatment of DC values is expected to be worthwhile [20]. The previous DC value thus needs to be stored, and the DC difference calculated, as visualized in Fig. 6.8. Signal *reset_dc_prev_reg* from the FSM resets the register that holds the previous DC value at startup to ensure first subtraction with 0. If the AC value or the DC difference consists of a negative number (which can be derived from the Most Significant Bit (MSB) of *value*), it needs to be decremented by 1, which is also performed in the Huffman calculator. Because of this required decrementing step of negative two's complement numbers, positive and negative numbers will be represented by their exact inverse binary format.

In the last step, the magnitude *category* for the calculated *coefficient* has to be selected for both DC and AC coefficients. The category selection can be achieved by a small combinational circuit instead of a large lookup table with decoder, as suggested by [11]. The implementation of the category selection in the current JPEG encoder only uses 2-input logic gates and is optimized to have a small pipeline depth and to have an as low as possible amount of gates to reduce the energy consumption.

Figure 6.9 shows the category selection block which converts the 12-bit coefficient in four pipeline stages to the 4-bit category which represents the magnitude of the coefficient. The XNOR gates in the light gray shaded area in the beginning make use of the fact that positive and negative binary numbers are their exact inverse, since the output of this XNOR step will be identical for a positive number x as

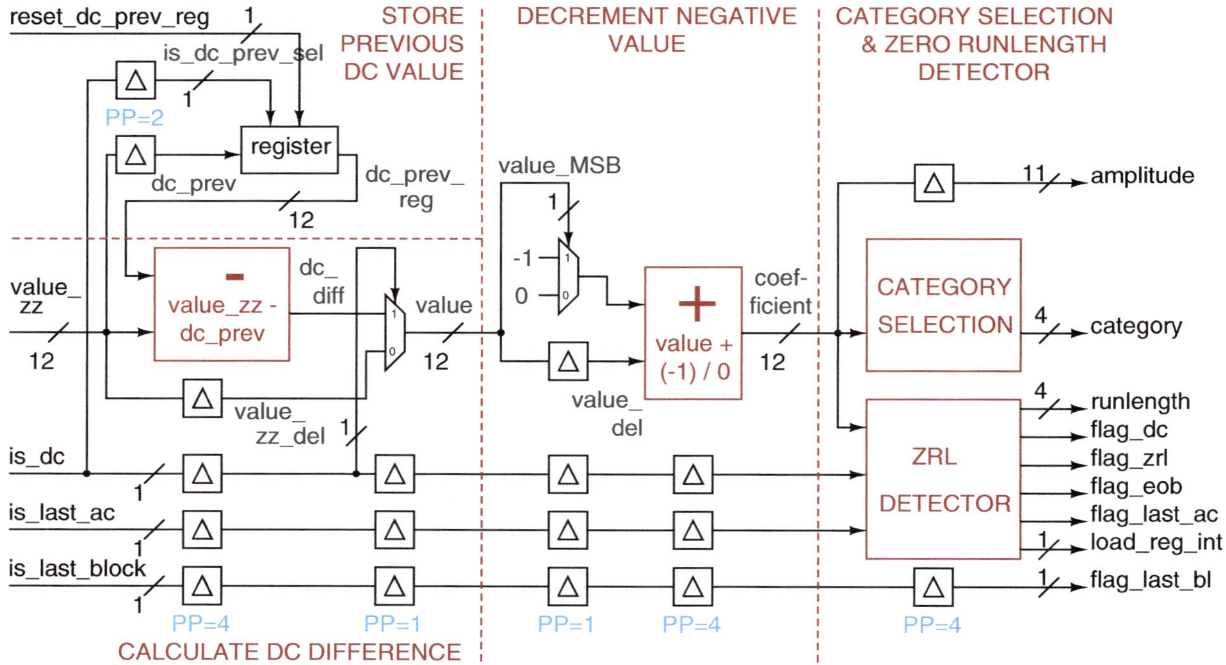

Fig. 6.8 Implementation of the Huffman calculator block of the Huffman encoder. The Δ symbol indicates a delay line of x latches with x given as $\mathrm{PP} = x$

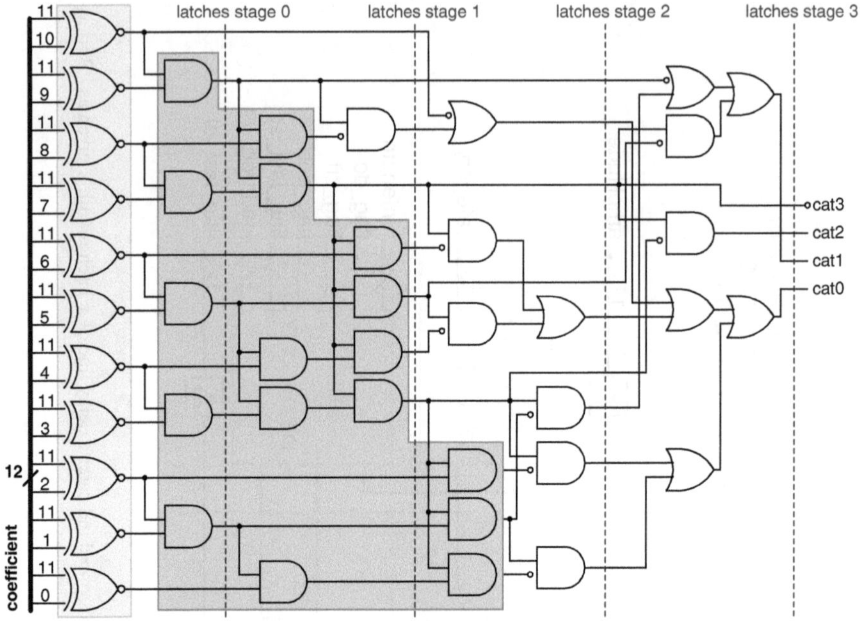

Fig. 6.9 Implementation of the category selection block inside the Huffman calculator

for its negative equivalent $-x$. The dark gray shaded area represents in fact a very long chain of AND gates, to systematically perform an AND operation on all output bits of the XNOR step. This AND chain detects the position of the highest first significant bit of the coefficient, which is important to determine the category of the coefficient. However, a chain of ten AND gates becomes a long path. Therefore, this has been optimized to the AND structure in the dark gray shaded area which performs the same operation with significantly less delay at the penalty of a higher gate count. The remaining logic gates of the category selection block are used to detect the exact value of the coefficient.

Note that each pipeline stage in this specific block only contains a maximum of two consecutive logic gates. The motivation for this is that the Zero RunLength (ZRL) detector block in parallel requires four pipeline stages. For timing variability reasons, it is thus better to balance the eight consecutive logic gates of the category selection block in the same amount of pipeline stages as the ZRL detector, instead of calculating them with maximum logic depth in a stage less. In Fig. 6.9, latches are inserted at the intersections of signal lines and the dashed latch lines. In general, when a block requires a certain pipeline depth, the pipeline stages of blocks in parallel have been balanced as much as possible throughout the entire design.

The Huffman encoder aggregates zero coefficients into runs of zeros (called *runlength*) to increase coding efficiency. A runlength is then combined with the magnitude category of the non-zero coefficient which terminates the run to determine the correct Huffman code [15]. The ZRL detector measures the runlength

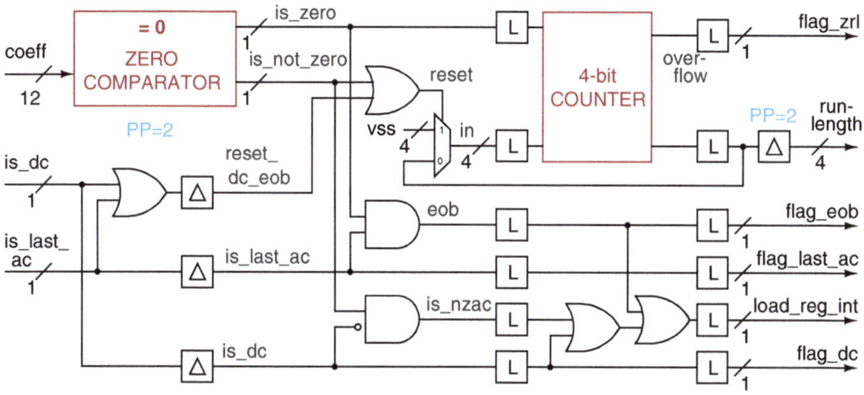

Fig. 6.10 Implementation of the ZRL detector block inside the Huffman calculator

by counting the number of zeros preceding a non-zero AC coefficient, as visible in Fig. 6.10. This block also detects two special codes, i.e. ZRL (*flag_zrl*) and EOB (*flag_eob*). The ZRL code is used for the rare case in which the run of zeros is very long. ZRL codes a run of 16 zeros, while End-Of-Block (EOB) codes an end-of-block condition which is inserted when the remaining AC coefficients are all 0. In the rare case in which the 63rd coefficient is not 0, EOB is not coded.

The *amplitude* output of the Huffman calculator (Fig. 6.8) consists of 11-bit since the MSB can be discarded after the step of decrementing negative values by 1.

6.4.4.3 Valid Checker

Only valid signals are Huffman encoded, i.e. DC, ZRL, EOB or non-zero AC values. The zero-valued AC coefficients should therefore be removed from the stream of coefficients. The valid checker block is shown in Fig. 6.11. It determines whether a coefficient is *valid*. The *load* signal in this block is thus only set to high whenever such a valid coefficient arrives. As can be seen in Fig. 6.10, part of the load logic (signal *load_reg_int*) has been efficiently inserted in the ZRL detector in the previous Huffman calculator block because an extra pipeline stage due to too many cascaded gates could thereby be avoided.

The valid checker consists of a set of four consecutive registers in which only valid coefficients are loaded. The registers save all necessary information for the following table access block. The output *valid* signal is only set to high during 1 clock period, which explains the different implementation without feedback of the 4th valid register, compared to the other three valid registers.

As required by the Huffman encoding algorithm, no ZRL symbols may directly precede an EOB symbol. Since the Huffman calculator will only add an EOB symbol at the position of the last, 63rd coefficient, it is possible that a maximum of 3 ZRL symbols are directly preceding the EOB. Therefore, they should be removed,

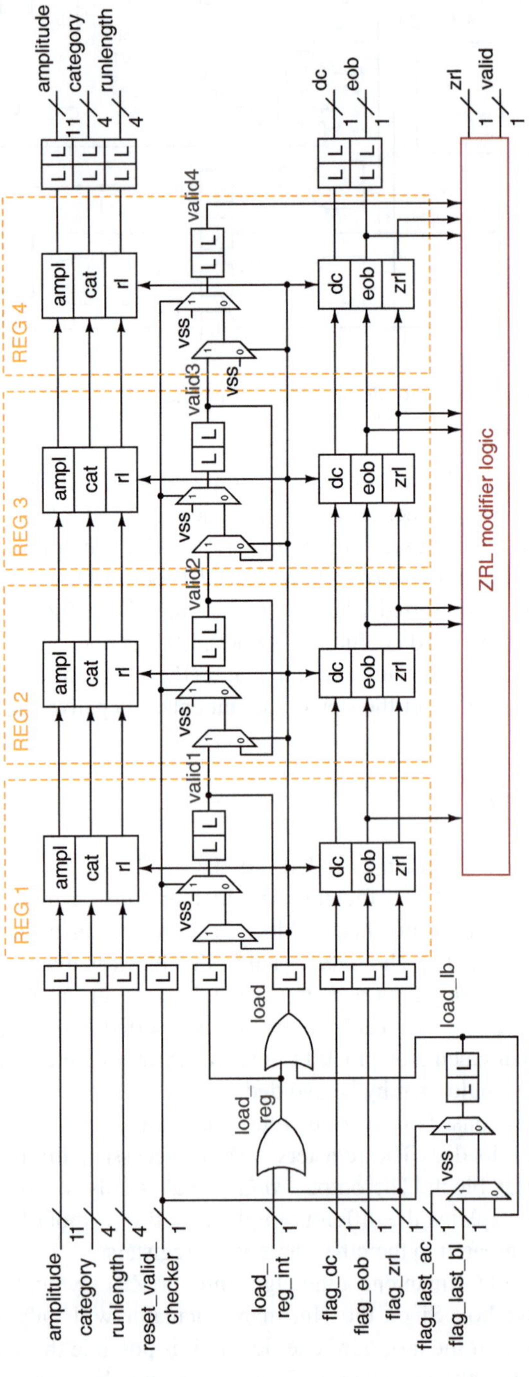

Fig. 6.11 Implementation of the valid checker block of the Huffman encoder

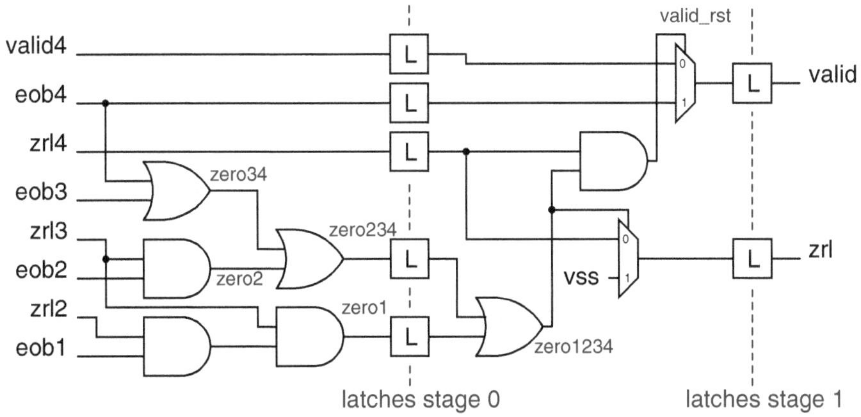

Fig. 6.12 Implementation of the ZRL modifier logic inside the valid checker

which is why there are four registers. The ZRL modifier logic performs this task, its implementation is shown in Fig. 6.12. The ZRL modifier block ensures that a ZRL signal is reset whenever there is an EOB symbol directly following one or more consecutive ZRL symbols. It lets all other ZRL symbols pass. Because this logic requires a pipeline depth of 2, the other register outputs need to be delayed as well, as can be seen in Fig. 6.11.

The *load* signal will remain high when the last AC coefficient of the very last 8×8 block of the image arrives at the valid checker (*load_lb*), because otherwise the valid data in registers $1-3$ will never get clocked through. By using *load_reg* as input signal for the 1st valid register, the output *valid* signal will only remain high until the last AC coefficient is clocked through. The *reset_valid_checker* signal from the FSM resets the *valid* and *load_lb* registers at startup.

6.4.4.4 Table Access

Figure 6.13 visualizes the implementation of the table access block. Its task is to retrieve the correct *ehufco* and *ehufsi* codes from the DC and AC Huffman tables. In the DC case, only the 4-bit category is needed, while for the AC case, both the category and the 4-bit runlength are necessary to determine the correct codes. Regular decoders are used to fetch data from the DC and AC tables. To minimize the switching energy of the tables, the input of their decoders is gated when the table is not accessed. The input gating is performed by adding logic gates which ensure that a non-existent entry of the decoders is then accessed by *cat_dc* or the combination of *cat_ac* and *rl_ac*.

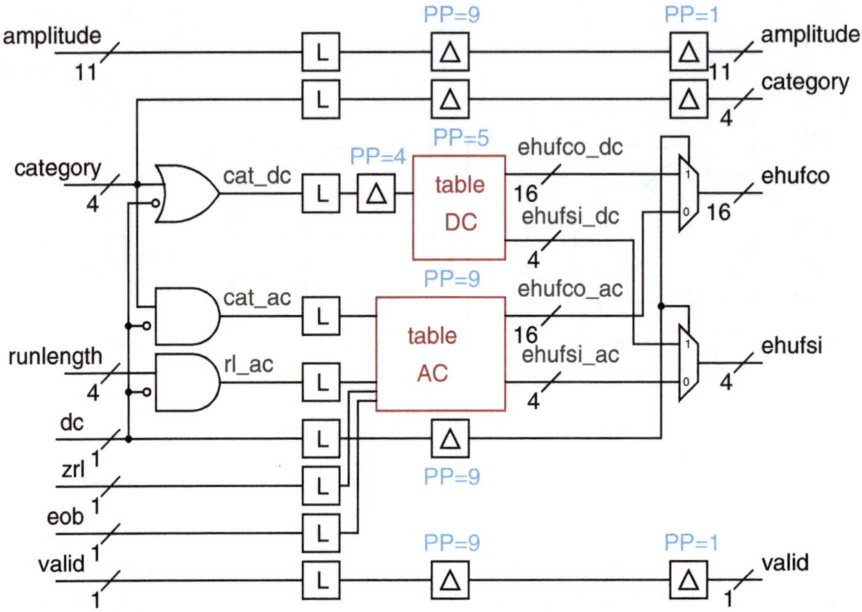

Fig. 6.13 Implementation of the table access block of the Huffman encoder

Table 6.1 JPEG lookup table specifications

	# Entries	# Bits/entry
Quantization table	64	12
DC Huffman table	12	20
AC Huffman table	162	20

6.4.5 Lookup Tables

There are three lookup tables used in the entire JPEG encoder: the quantization, DC and AC Huffman tables. The specifications of the different tables can be found in Table 6.1. Figure 6.14 shows the table implementation: a table consists of a decoder, a register matrix and entry selectors. As mentioned before, the Huffman tables use a full address decoder for which they receive an address from the Huffman encoder, while the quantization table can be implemented with an energy-efficient one-hot decoder. The output of the decoders consists of *n* word lines wl_i which are used as inputs for the entry selectors. Latches are inserted at the word line outputs of the full address decoders, to limit the number of required latches in the entry selectors inside the register matrix. These extra latches are not necessary for the one-hot decoder because it consists solely of a register loop.

The tables are implemented as register matrices. They are not implemented as SRAM memories because the energy consumption of SRAMs is dominated by standby leakage, rather than by dynamic energy. This stands in contrast to the

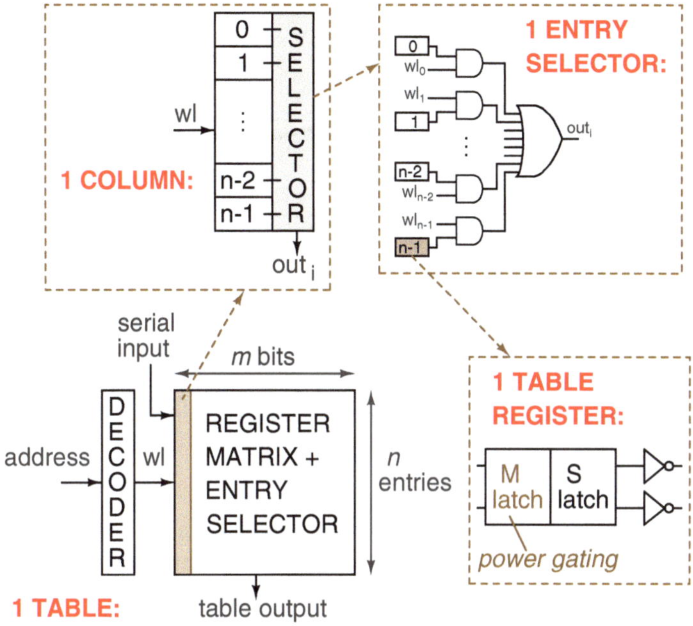

Fig. 6.14 Implementation of the lookup tables

main argument for working in the ultra-low-voltage region, which is reducing the dynamic energy consumption. The most limiting factor of sub- or near-threshold SRAM is that its speed is very low [4], and would be, in fact, too low for this design, as will be seen from the measurement results later in Sect. 6.5. This is the primary reason why register matrices, which do enable sufficient speed at the same supply voltage of this design, are used. Moreover, for the required number of bits, the peripheral area and energy overhead of SRAM would be too high. Furthermore, standard SRAM requires ratioed logic, which is undesirable due to the high sensitivity to variations of ultra-low-voltage circuits. If a second voltage domain would be allowed, lookup table improvements would be possible, as will be discussed in Sect. 6.7.

The register matrix is serially written at startup and is from then on only accessed for reading words. At startup, the data is serially clocked in into the register matrix, as visible in Fig. 6.14. After startup, the slave (S) latch of the table registers stores the data, while the master (M) latch is only necessary during the serial clocking in. To significantly reduce the leakage of the register matrices, the master latches are therefore power gated. Their ground supply is the same as the overall ground V_{ss}, but their power supply $V_{dd,master}$ is a separate supply which is pulled to ground after startup.

The implementation of the table registers is also shown in Fig. 6.14. The latches have the same fully differential topology as all the latches in this design. However, the table registers are different compared to the other registers: inverters are added

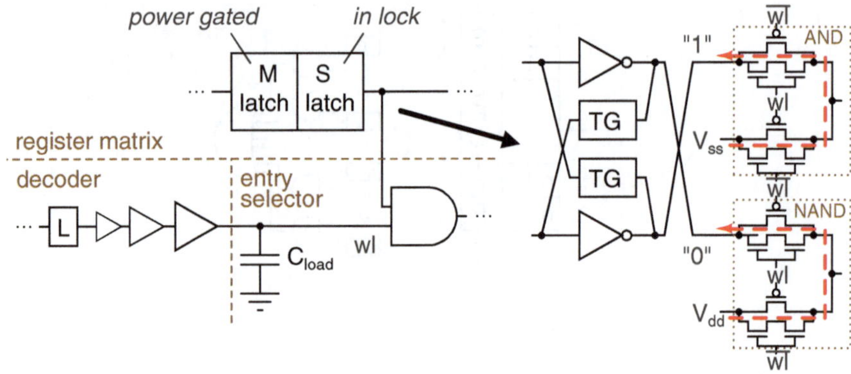

Fig. 6.15 Visualization of sneak leakage path problem with table register. Inverters are added at the outputs of the slave latch to overcome this

at both complementary outputs of the slave latch. Because of the large load for the word lines, both in terms of capacitance due to the high amount of logic gates attached to them and in terms of wire capacitance, they were buffered in the decoder. However, when applying intra-die variations through MC simulations, in a few cases the data stored in the slave latches became compromised, as visualized in Fig. 6.15. This problem was due to the fact that the complementary word line signals were sometimes not perfectly differential due to mismatch. While this does not pose a problem in a regular setting without buffers since all logic gates are slow at ultra-low supply voltages, the buffers make the transition edges much sharper and therefore enlarge the very small $1 - 1$ overlap of the *wl* signals. In that case, sneak leakage paths in the AND gates in the entry selectors can compromise the outputs of the slave latch and even its saved state if that latch is weakened by intra-die variations. When one of their output levels is severely corrupted, the cross-coupled inverters will at some point flip and the table register will lose its saved value. Since the weakened slave latch is in lock, the cross-coupled inverters aggravate the situation considerably by suddenly flipping state.

One option to resolve this issue would be to lower the mismatch by upsizing the buffers and the latch before them in the decoder. However, the entire decoder (i.e. the logic and latches) would hence need to be upsized to be able to drive these upsized buffers. This brings a large cost in energy consumption, both in leakage and in dynamic energy. Therefore, inverters are added at the output signals of the slave latch so that even if the sneak leakage paths would interfere with the output of the inverters, the table register would never lose its saved data. Extensive MC simulations confirmed that this last option solved the issue. Because the data in the tables does not change anymore after startup, these inverters only add to leakage and do not consume switching energy during JPEG encoding operation. Moreover, the added leakage of the extra inverters is also partially compensated from 2 to 1.5 times more leakage by downsizing the inverters in the slave latch

($W_{nMOS} = W_{min}$ and $W_{pMOS} = 5 \cdot W_{min}$) because their necessary drive current was reduced. To conclude, the implementation with extra inverters is necessary to guarantee correct functionality.

The entry selector (see Fig. 6.14) has AND gates in its first stage, while the remaining stages are used to perform an n-input OR operation, implemented with a tree of 2-input OR gates. If the number of inputs is not a power of 2, a tree equalization technique is employed to guarantee a symmetrical OR tree: few dummy OR gates are inserted to ensure equal path delay.

Note that both the decoder and the entry selectors are implemented using the same methodology as the rest of the JPEG encoder, i.e. they are pipelined and a pipeline stage has maximally three consecutive Transmission Gate (TG) logic gates. However, because of the long wires in the entry selectors, the delay of those pipeline stages must be simulated so that it does not exceed the delay of a regular pipeline stage. Through parasitic extractions, the wire capacitances are determined. These are incorporated in a delay check method which compares both delays and adjusts a certain pipeline stage or inserts buffers where necessary.

6.5 Measurement Results

The JPEG encoder is fabricated in a 40 nm CMOS technology. To prove that the chip is fully functional, a raw micrograph of a die was taken. This image was split into 8 × 8 blocks and these were fed as an input to the chip. The chip then JPEG encoded the raw image. Figure 6.16 thus shows a JPEG image of the chip which was encoded through the chip itself.

Figure 6.16 also highlights the various building blocks and their respective active areas. The total active area of the JPEG encoder is 0.557 mm^2. The dense layout of the JPEG encoder is carried out using DPG, as explained in Sect. 4.4.2. An exception is the layout of the three tables, which was done manually because their regularity allowed an optimized structure. Table 6.2 gives the pipeline depths of the different subblocks, as well as the total pipeline depth of the JPEG encoder.

Measurements on the JPEG encoder were performed on a total of 26 dies, in order to be able to adequately study the variations. Figure 6.17 provides the distribution of the measured minimal supply values at which the dies were still functional. Measurement results show that the $V_{dd,min}$ of the JPEG encoder is 210 mV, which is lower than the expected 230 mV, as discussed before in Sect. 6.3. However, the mean value μ of $V_{dd,min}$ out of the 26 dies is 232 mV with a standard deviation σ of 12.2 mV.

The upper plot of Fig. 6.18 shows a boxplot of the measured maximum operating frequency as function of V_{dd}. At the minimal supply of 210 mV, a clock frequency of 5 MHz is achieved, as visible in Fig. 6.19, which zooms in on the ultra-low-voltage region. The targeted speed of at least tens of MHz has thus been obtained. Frequencies of up to 275 MHz are possible with supplies from 210 to 550 mV. Frequencies of 25, 50 or 100 MHz are achieved at supplies of 300, 350 and 410 mV,

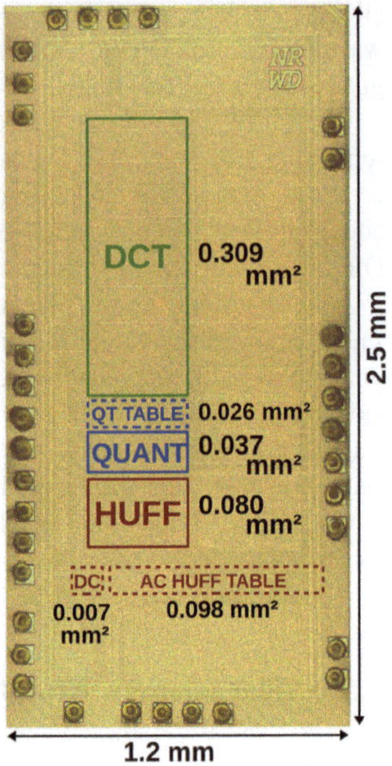

Fig. 6.16 Encoded JPEG image of the chip. Die size is $1.2 \times 2.5\,\text{mm}^2$ [17]

Table 6.2 Overview of pipeline depths of different subblocks of the JPEG encoder [17]

	Pipeline depth
2D-DCT	458
Quantization	22
Zigzag	130
Huffman coding	36
Total JPEG encoder	646

respectively. In the visible supply range of Fig. 6.18, the percentage variation σ/μ in operating frequency of the 26 measured dies is only 8.6 %.

The upper plot of Fig. 6.18 shows the measured energy consumption per pixel as function of V_{dd}. Overall, the JPEG encoder achieves an energy consumption of less than 50 pJ/pixel for clock frequencies below 275 MHz. The MEP occurs at a supply of 330 mV. The chip then consumes 29.01 pJ/pixel at an operating frequency of 41 MHz (see Fig. 6.19). Naturally, the highest energy-efficiency is obtained at the Minimum-Energy Point (MEP), but interesting is that the region around the MEP is quite flat. For supplies from 290 to 350 mV, the energy consumption stays below 30 pJ. Depending on the desired operating frequency, a high energy-efficiency can thus still be obtained in a relatively large region around the MEP. Across the depicted supply range in Fig. 6.18, the percentage variation in energy consumption per pixel is 5.4 %, demonstrating the variation-resilience of the total design.

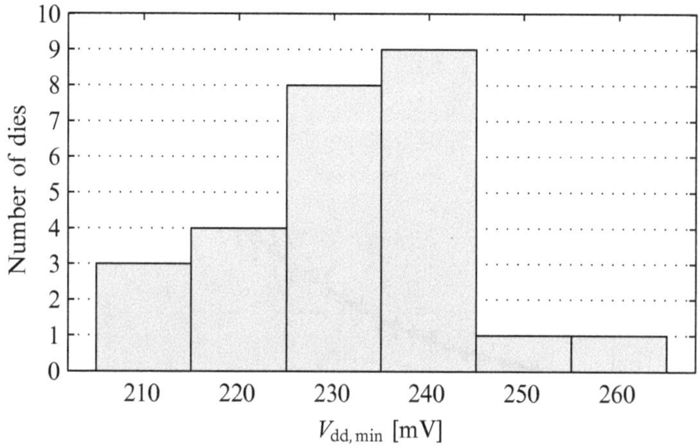

Fig. 6.17 Distribution of the minimal functional supply voltage $V_{dd,min}$ of the 26 measured dies

The division of the total energy consumption at the MEP is provided in Fig. 6.20, as well as the contribution of leakage to the different subblocks. The timing block, which consist of the non-overlapping clock generator and the clock tree, consumes 58.8 % of the energy. The three other building blocks combined consume less than half of the total energy. The contribution of leakage to each block's energy consumption can be seen as well. Although design effort was made to reduce the leakage of the tables as much as possible, it is apparent that the quantization and zigzag & Huffman blocks have a much higher percentage of leakage than the 2D-DCT does.

This can also be seen in Fig. 6.21, which visualizes the contribution of leakage to the total energy as function of V_{dd} for the different subblocks, as well as for the entire JPEG encoder. The timing block has the lowest percentage of leakage, because the switching energy of the clock tree is very high. At the MEP, leakage accounts for 40 % of the total energy. Observe that the register tables contribute significantly to leakage, as the quantization and zigzag & Huffman blocks have a much higher contribution of leakage than the 2D-DCT, which does not contain a table. Section 6.7 will elaborate on which measures could be taken in the future to reduce the leakage of the lookup tables.

Figure 6.22 presents a boxplot of the measured EDP as function of V_{dd}. At the MEP, the Energy-Delay Product (EDP) is 0.716 pJ.μs.

6.6 State-of-the-Art Comparison

Table 6.3 provides a state-of-the-art comparison between this work and the only other published ultra-low-voltage JPEG encoder [16], which was fabricated in a 65 nm CMOS technology. In [16], the 2D-DCT and quantization are joined in

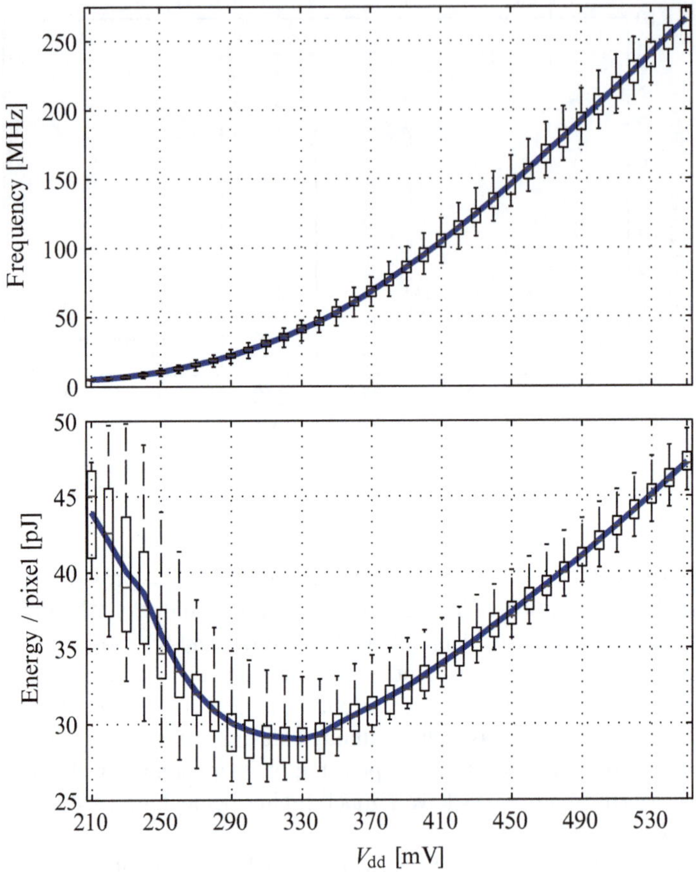

Fig. 6.18 Boxplot of the measured maximum clock frequency and energy consumption per operation as function of V_{dd}

a so-called engine and such an engine is operating at 1 voltage domain, while the Huffman encoder is operating in a 2nd higher voltage domain. The pipelined architecture consists of four parallel engines and a single Huffman encoder which runs at 4× the engine clock.

This work is able to function at a minimum supply of 210 mV, while [16] is only able to reach a $V_{dd,min}$ of 400 mV in the engine. At 400 mV, the engines are able to operate at a clock frequency of 2.5 MHz and the Huffman encoder runs at 10 MHz at 600 mV. This work achieves a throughput frequency of 5 MHz at the minimum supply, and 41 MHz at the MEP, significantly outperforming the throughput numbers of [16]. Unfortunately, a direct comparison between the energy consumptions cannot be made, because [16] only provides the average energy consumption per pipeline stage. Since the total number of pipeline stages is not mentioned in [16], it is not possible to calculate and compare the total energy consumption, nor the EDP.

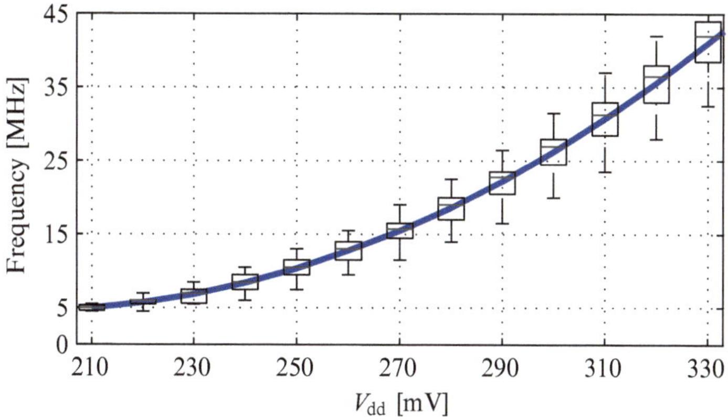

Fig. 6.19 Zoomed-in boxplot of the measured maximum clock frequency as function of V_{dd}

Fig. 6.20 Energy division at the MEP (V_{dd} = 330 mV). The contribution of leakage to each different subblock is indicated with a white circle

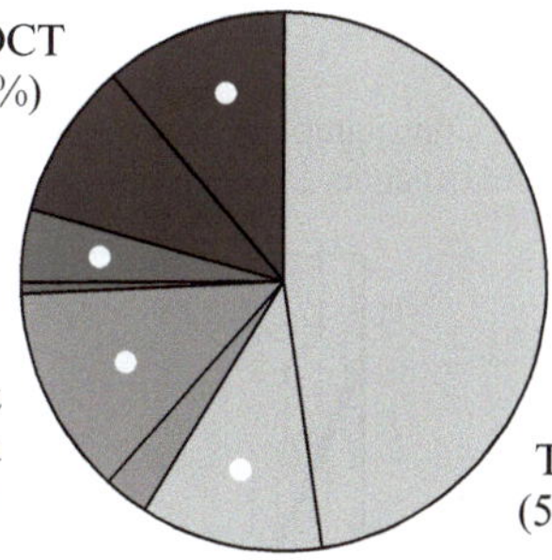

Nonetheless, because the performed design efforts of the JPEG encoder are generally applicable to achieve a high energy-efficiency and variation-resilience for ultra-low-voltage designs, an as extensive state-of-the-art comparison of such designs as possible is provided next. Although energy consumptions of different large ultra-low-voltage designs are difficult to compare, their reported operating frequencies can be compared. Hence, this allows a much broader state-of-the-art comparison. Therefore, Fig. 6.23 shows a comprehensive state-of-the-art frequency comparison among all previously published designs (to the author's knowledge) fabricated in a 65 nm CMOS technology or smaller, which were able to function at a supply voltage of 500 mV or lower. The present work exceeds the speed performance of previous ultra-low-voltage designs in advanced nanometer technologies. Only [7] achieve a similar frequency.

The measurements of the JPEG encoder have been carried out at room temperature. The results of Fig. 6.23 are all measured operating frequencies at room temperature as well, except for the results of [9], which were obtained at 50 °C. As demonstrated in Fig. 2.12, the delay of an ultra-low-voltage circuit decreases at higher temperatures. Therefore, the reported frequencies of [9] are somewhat biased since they are expected to become lower at room temperature.

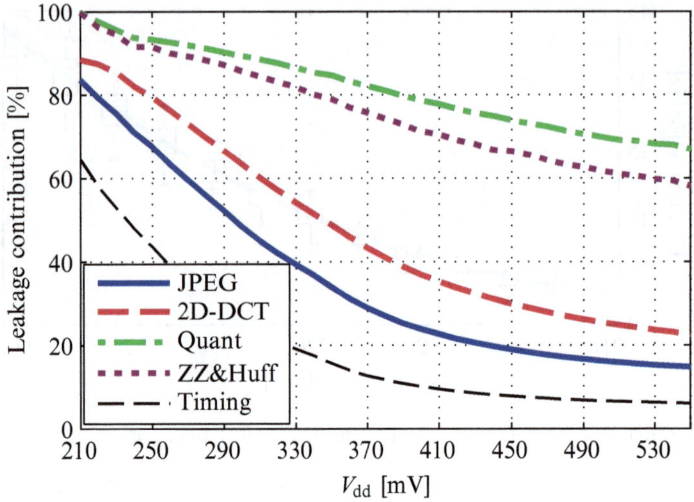

Fig. 6.21 Contribution of leakage to the total energy per subblock individually as well as for the entire JPEG encoder

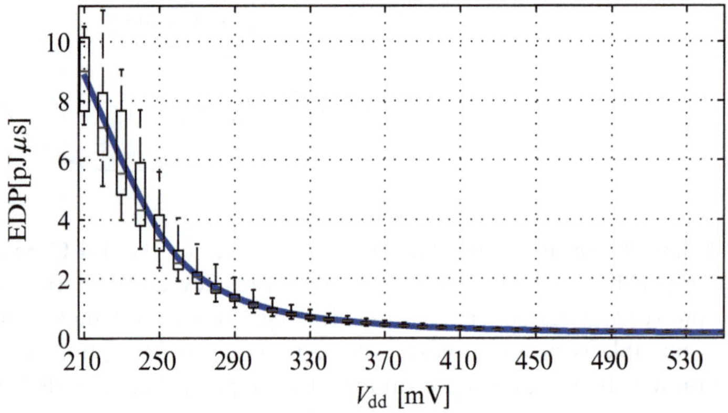

Fig. 6.22 Boxplot of the measured EDP as function of V_{dd}

To verify that this JPEG encoder achieves state-of-the-art variation-resilience, Table 6.4 compares the results of this work for both frequency and energy to the designs of Fig. 6.23 that reported variation numbers, at the reported supplies. Unfortunately, only three of those references report these variation numbers. References [7, 12, 13] all consist of 65 nm CMOS technologies, while this work has been designed in a 40 nm CMOS technology which is expected to have increased variability.

Table 6.3 State-of-the-art comparison of ultra-low-voltage JPEG encoders [17]

		This work	[16]
CMOS technology		40 nm	65 nm
Active area	[mm^2]	0.557	1.960
$V_{dd,min}$	[mV]	210	400 (engine)/600 (Huffman)
Voltage @ MEP	[mV]	330	400/600
Frequency @ MEP	[MHz]	41.0	2.5/10.0
Energy/cycle @ MEP *	[pJ]	0.045	3.0 (for four engines) + 1.8
Energy/block @ MEP	[pJ]	1,857	–
Energy/pixel @ MEP	[pJ]	29.01	–
EDP @ MEP	[pJ.μs]	0.716	–

*This is design-dependent, but it is the only energy figure provided in [16]

Table 6.4 State-of-the-art comparison of reported variation numbers of CMOS designs from Fig. 6.23. The relative variation percentage σ/μ of frequency and energy is compared at the same V_{dd}

V_{dd}	σ/μ	References & technology		This work
300 mV		[7], 65 nm		
	Frequency	7 %		10.5 %
	Energy	2 %		7.2 %
500 mV		[13], 65 nm	[12], 65 nm	
	Frequency	7.5 %	13.3 %	5.4 %
	Energy	–	9.0 %	2.4 %

6.7 Lookup Table Improvements

In Figs. 6.20 and 6.21, it can be seen that a large portion of the used energy is caused not by active switching but rather by leakage. The quantization and the zigzag & Huffman encoder blocks both have a share of leakage energy of 60 % or more, even at the highest speed. These are the two blocks that contain lookup tables. These tables have, due to their nature, a very low activity. Nevertheless, they are designed using the same methodology as the rest of the JPEG encoder, resulting in very large register banks. Reducing the leakage energy of these circuits is thus of crucial importance to reach an even more energy-efficient design. As explained earlier, register tables have been used to be able to operate in a single supply and clock domain, because an SRAM operating at such low supply voltages does not reach sufficient speed to allow this.

The solution lies in revisiting the reasoning that led to the adoption of near-threshold design [18]. As explained in Chap. 1, ultra-low-voltage design is attractive for systems with high activities, where reducing dynamic energy has a significant impact on the total energy consumption. This can be achieved by decreasing the supply voltage as much as possible. By using LVT transistors, the lowest delay at a certain supply voltage is achieved. To reach sufficient speed, LVT transistors are the most attractive to use for ultra-low-voltage designs, as discussed in Sect. 2.4. However, in the case of the lookup tables of this JPEG encoder, and of memories in general, not dynamic energy but leakage is dominant. The most effective manner to

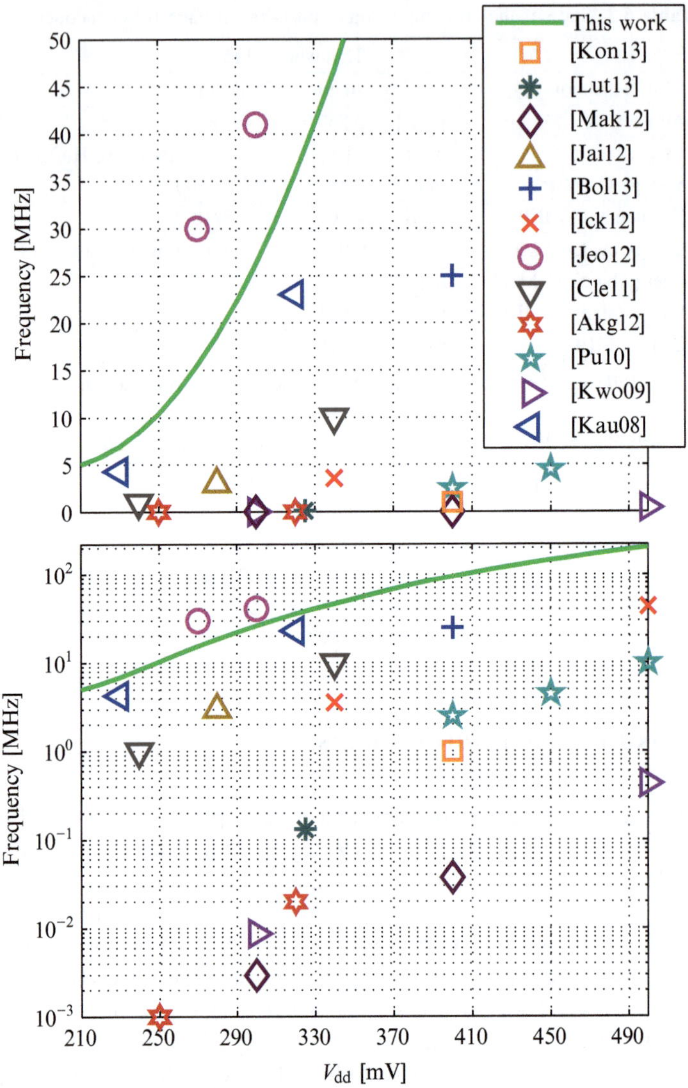

Fig. 6.23 State-of-the-art frequency comparison among all other previously published ultra-low-voltage designs in advanced nanometer CMOS technologies. All frequencies were measured at room temperature, except for those of [9] which were measured at 50 °C. The lower plot is the logarithmic version, to better visualize the smaller frequencies

reduce leakage is to use HVT transistors. The supply voltage V_{dd} can then be altered to achieve an energy-efficient memory at the wanted speed.

To illustrate this, Fig. 6.24 shows the mean leakage, the worst-case read current and the stability of a classic 6T-SRAM cell as function of the supply voltage for three different V_T-options of transistors. It is clear that for the same read current

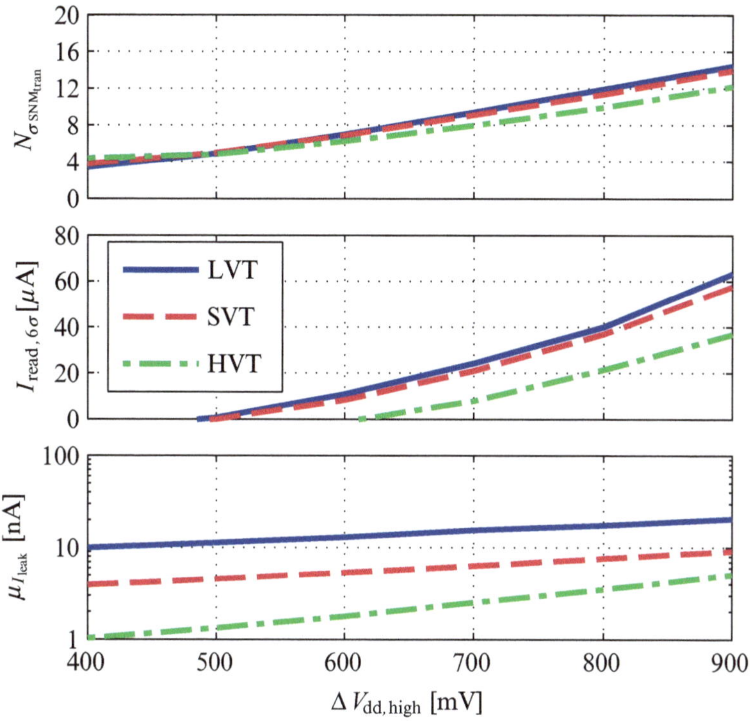

Fig. 6.24 6T SRAM cell parameters as function of supply and threshold voltage [18]

(and thus speed), increasing both V_T and V_{dd} is beneficial both for leakage and for stability [19]. As an example, one can compare a cell of LVT transistors operating at a supply of 500 mV with a HVT cell at 700 mV. This has the following results: the leakage (lower plot) is reduced significantly, while a higher speed (middle plot) can be obtained at a higher stability (upper plot). To conclude, energy-efficient design is not simply equal to ultra-low-voltage design because activity plays an important role. In the case of SRAM, the energy-efficiency is higher with a higher supply voltage $V_{dd,high}$ and with HVT devices for the cells.

Aside from combining higher supply voltages with HVT transistors for the memory cells, other circuit techniques can be used to reduce the active energy or leakage power of a memory as well. For instance, word lines can be divided into local word lines [21], that only activate the word to be read or written. This highly reduces wasteful activity. Furthermore, most active energy in a memory is used by the data transfers on the bit lines. These bit lines are highly capacitive due to the large number of connected cells. Dividing these bit lines in local bit lines connected to a single global bit line reduces this load significantly [8]. The energy use on these lines can be further reduced by using low swing signals. When employing the memory in a near-threshold design, the supply of the near-threshold circuit might be reused for this.

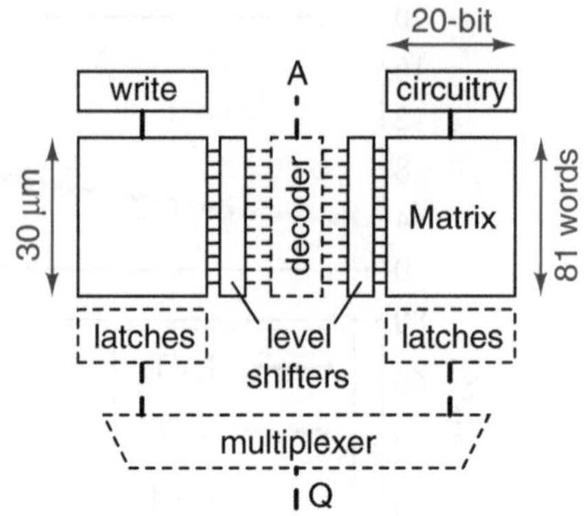

Fig. 6.25 Proposed architecture of an SRAM lookup table [18]

Although energy-efficient SRAM design is not a topic of this book, it is interesting to observe what impact it would have to substitute the register tables which are operating at the same supply voltage as the rest of the JPEG encoder by an energy-efficient SRAM functioning at a higher supply $V_{dd,high}$. This supply voltage would then be chosen to achieve the same speed as the JPEG encoder. As a case study, the design of the AC Huffman table with 162 words of 20-bit is considered. With only 3,240 bits, this is a very small memory. 6T SRAM cells with HVT transistors can be used with a matrix voltage $V_{dd,high}$ of 700 mV. This results in a read current sufficiently high to reach the 41 MHz MEP speed and in sufficient stability (see Fig. 6.24).

Figure 6.25 shows the proposed architecture of the AC Huffman table. To shorten the bit lines, each row contains two words. The decoder is placed in the middle of the matrix. Bit lines can be divided in ten blocks of eight words and a single extra word. As these tables are not written much, the local bit lines can be connected with the global bit line with a simple pass transistor. The 330 mV V_{dd} supply of the JPEG encoder is then used as a low swing voltage for read operation. Writing is done with full swing signaling, but this has no influence on energy as it is only done during startup.

The most active part, i.e. the decoder, has V_{dd} as supply (indicated by the dashed borders) and is implemented using the same design style as the rest of the JPEG encoder. Level shifters are used to shift the output of the decoder to $V_{dd,high}$ (indicated by the solid borders). As V_{dd} is used as the low swing voltage on the bit lines, these have the same logic levels as the JPEG encoder. No sense amplifiers or level shifters are needed after the matrix and the output can simply be captured by latches working at V_{dd}.

It is clear that when allowing a second, higher supply voltage in the JPEG encoder, superior memory architectures (such as Fig. 6.25) could be employed, yielding large reductions in terms of area and leakage. For example, the cell area and the cell leakage power then both decrease significantly from 13.505 to 0.475 μm^2 and from 6.86 to 1.73 nW, respectively. An SRAM implementation at a higher supply voltage and the same speed could result in a more energy-efficient design, provided that the energy cost of the supply generation would not be too high.

6.8 Conclusion

This chapter concluded the presentation of the different prototypes realized in this book. As a fourth and final prototype, a full JPEG encoder has been implemented in a 40 nm CMOS technology. This design incorporated both the gate-level and architecture-level strategy discussed in Chaps. 3 and 4 as well as the insights which were obtained during the design and measurements of the previous three prototypes of Chap. 5. Throughout the entire JPEG encoder, increasing energy-efficiency has been the key factor for all design decisions, as extensively discussed in this chapter. The chip achieved state-of-the-art speed and energy numbers for ultra-low-voltage operation, and demonstrated a high variation-resilience. To conclude, the described design efforts are thus effectively validated and it is shown that they are generally applicable for any ultra-low-voltage DSP design.

References

1. Akgun O, Rodrigues J, Leblebici Y, Owall V (2012) High-level energy estimation in the sub-Vt domain: Simulation and measurement of a cardiac event detector. IEEE Trans Biomed Circuits Syst 6(1)15–27. DOI: 10.1109/TBCAS.2011.2157505
2. Bol D, De Vos J, Hocquet C, Botman F, Durvaux F, Boyd S, Flandre D, Legat JD (2013) Sleepwalker: A 25-MHz 0.4-V sub-mm^2 7-μW/MHz microcontroller in 65-nm LP/GP CMOS for low-carbon wireless sensor nodes. IEEE J Solid-State Circuits 48(1):20–32. DOI: 10.1109/JSSC.2012.2218067
3. Clerc S, Abouzeid F, Argoud F, Kumar A, Kumar R, Roche P (2011) A 240mV 1MHz, 340mV 10MHz, 40nm CMOS, 252 bits frame decoder using ultra-low voltage circuit design platform. In: Proceedings of the IEEE international conference on electronics, circuits and systems (ICECS), pp 117–120. DOI: 10.1109/ICECS.2011.6122228
4. Cosemans S, Dehaene W, Catthoor F (2008) A 3.6pJ/access 480MHz, 128Kbit on-chip SRAM with 850MHz boost mode in 90nm CMOS with tunable sense amplifiers to cope with variability. In: Proceedings of the IEEE European solid-state circuits conference (ESSCIRC), pp 278–281. DOI: 10.1109/ESSCIRC.2008.4681846
5. Ickes N, Gammie G, Sinangil M, Rithe R, Gu J, Wang A, Mair H, Datla S, Rong B, Honnavara-Prasad S, Ho L, Baldwin G, Buss D, Chandrakasan A, Ko U (2012) A 28 nm 0.6 V low power DSP for mobile applications. IEEE J Solid-State Circuits 47(1):35–46. DOI: 10.1109/JSSC.2011.2169689
6. Jain S, Khare S, Yada S, Ambili V, Salihundam P, Ramani S, Muthukumar S, Srinivasan M, Kumar A, Gb S, Ramanarayanan R, Erraguntla V, Howard J, Vangal S, Dighe S, Ruhl G, Aseron P, Wilson H, Borkar N, De V, Borkar S (2012) A 280mV-to-1.2V wide-operating-range IA-32 processor in 32nm CMOS. In: Proceedings of the IEEE international solid-state circuits conference (ISSCC), pp 66–68. DOI: 10.1109/ISSCC.2012.6176932
7. Jeon D, Seok M, Chakrabarti C, Blaauw D, Sylvester D (2012) A super-pipelined energy efficient subthreshold 240 MS/s FFT core in 65nm CMOS. IEEE J Solid-State Circuits 47(1):23–34. DOI: 10.1109/JSSC.2011.2169311
8. Karandikar A, Parhi K (1998) Low power SRAM design using hierarchical divided bit-line approach. In: Proceedings of the IEEE international conference on computer design (ICCD), pp 82–88. DOI: 10.1109/ICCD.1998.727027

9. Kaul H, Anders M, Mathew S, Hsu S, Agarwal A, Krishnamurthy R, Borkar S (2008) A 320mV 56μW 411GOPS/Watt ultra-low voltage motion estimation accelerator in 65nm CMOS. In: Proceedings of the IEEE international solid-state circuits conference (ISSCC), pp 316–317. DOI: 10.1109/ISSCC.2008.4523184
10. Konijnenburg M, Cho Y, Ashouei M, Gemmeke T, Kim C, Hulzink J, Stuyt J, Jung M, Huisken J, Ryu S, Kim J, de Groot H (2013) Reliable and energy-efficient 1MHz 0.4V dynamically reconfigurable SoC for ExG applications in 40nm LP CMOS. In: Proceedings of the IEEE international solid-state circuits conference (ISSCC), pp 430–431. DOI: 10.1109/ISSCC.2013.6487801
11. Kovac M, Ranganathan N (1995) JAGUAR: a fully pipelined VLSI architecture for JPEG image compression standard. Proceedings of the IEEE 83(2):247–258. DOI: 10.1109/5.364464
12. Kwong J, Ramadass Y, Verma N, Chandrakasan A (2009) A 65 nm sub-Vt microcontroller with integrated SRAM and switched capacitor DC-DC converter. IEEE J Solid-State Circuits 44(1):115–126. DOI: 10.1109/JSSC.2008.2007160
13. Lutkemeier S, Jungeblut T, Berge H, Aunet S, Porrmann M, Ruckert U (2013) A 65nm 32b subthreshold processor with 9T multi-Vt SRAM and adaptive supply voltage control. IEEE J Solid-State Circuits 48(1):8–19. DOI: 10.1109/JSSC.2012.2220671
14. Makipää J, Turnquist MJ, Laulainen E, Koskinen L (2012) Timing-error detection design considerations in subthreshold: An 8-bit microprocessor in 65nm CMOS. J Low Power Electron Appl 2(2):180–196. DOI: 10.3390/jlpea2020180
15. Pennebaker WB, Mitchell JL (1993) JPEG: Still image data compression standard. Kluwer Academic Publishers, Dordrecht
16. Pu Y, Pineda de Gyvez J, Corporaal H, Ha Y (2010) An ultra-low-energy multi-standard JPEG co-processor in 65 nm CMOS with sub/near threshold supply voltage. IEEE J Solid-State Circuits 45(3):668–680. DOI: 10.1109/JSSC.2009.2039684
17. Reynders N, Dehaene W (2014) A 210mV 5MHz variation-resilient near-threshold JPEG encoder in 40nm CMOS. In: Proceedings of the IEEE international solid-state circuits conference (ISSCC), pp 456–457
18. Reynders N, Rooseleer B, Dehaene W (2014) Energy-efficient logic and SRAM design: A case study. In: Proceedings of the IEEE faible tension faible consommation conference (FTFC), pp 1–4. DOI: 10.1109/FTFC.2014.6828616
19. Rooseleer B, Cosemans S, Dehaene W (2012) A 65 nm, 850 MHz, 256 kbit, 4.3 pJ/access, ultra low leakage power memory using dynamic cell stability and a dual swing data link. IEEE J Solid State Circuits 47(7):1784–1796. DOI: 10.1109/JSSC.2012.2191316
20. Wallace G (1992) The JPEG still picture compression standard. IEEE Tran Consum Electron 38(1):xviii–xxxiv. DOI: 10.1109/30.125072
21. Yoshimoto M, Anami K, Shinohara H, Yoshihara T, Takagi H, Nagao S, Kayano S, Nakano T (1983) A divided word-line structure in the static RAM and its application to a 64K full CMOS RAM. IEEE J Solid State Circuits 18(5):479–485. DOI: 10.1109/JSSC.1983.1051981

Chapter 7
Conclusion

The aim of this book was to develop circuit and architectural techniques to design ultra-low-voltage digital circuits with high energy-efficiency in CMOS technologies. Another essential target of this research has been to design circuits which are able to operate at frequencies of $n \times 10\,\text{MHz}$. Furthermore, it is found to be imperative to achieve a high variation-resilience of these circuits in order to guarantee a high yield. Techniques on how to achieve these goals have been demonstrated throughout the book.

This chapter will provide an extensive conclusion of this research. It includes an overview of the conclusions which were reached in Sect. 7.1, while the obtained results of the prototypes are situated in the current state-of-the-art in literature in Sect. 7.2. Section 7.3 presents the main contributions of this research. Finally, Sect. 7.4 concludes this book by a look at future perspectives for follow-up research.

7.1 General Conclusions

A summary of the conclusions of the different chapters in this book is now given, to recapitulate the insights which have been gained throughout this work.

Chapter 1 introduced the topic of this research in a larger context: why is today's society craving for more energy-efficient electronic devices and how can ultra-low-voltage digital circuits provide an answer to this need for ever increasing energy-efficiency? A brief history of scaling in general and of ultra-low-voltage digital research specifically has been given in this chapter. Different mechanisms play a role in the power and energy consumption of a system. Detailed insight in these mechanisms is therefore necessary to be able to realize highly energy-efficient systems. Furthermore, the range of applications which can greatly benefit from ultra-low-voltage operation is discussed: it consists of applications

N. Reynders, W. Dehaene, *Ultra-Low-Voltage Design of Energy-Efficient Digital Circuits*, Analog Circuits and Signal Processing,
DOI 10.1007/978-3-319-16136-5_7

which are severely energy-constrained but have less stringent speed performance requirements. Lastly, an overview of the current state-of-the-art in literature is provided, so as to show the readers which improvements are still necessary to increase the industrial relevance of ultra-low-voltage research.

Chapter 2 focused on device-level behavior of transistors operating at ultra-low supply voltages by studying the fundamentals of sub-threshold operation. The exponential behavior of those transistors introduces quite some challenges to ultra-low-voltage design. All important challenges are discussed extensively: the inherently low to moderate performance of ultra-low-voltage circuits, the reduced current ratios which pose a threat to reliable functionality, the exponential sensitivity to both inter- and intra-die variations which compromises yield, as well as the impact of temperature on such circuits. It is essential to find solutions to cope with all these challenges to be able to successfully design ultra-low-voltage systems.

The two CMOS technologies which are used throughout this research are explored, as well as the impact of scaling on circuits operating in the ultra-low-voltage region. An equation to calculate the theoretical minimum as well as a practical minimum supply voltage for a specific technology is proposed. Furthermore, after thorough analysis, this chapter recommended the use of LVT transistors in the high-performance process of CMOS technologies, the reason being that these devices offer the highest available current for a certain supply voltage. Therefore, a maximal sub-threshold speed can be guaranteed.

Chapter 3 went a step higher on the abstraction level ladder by examining which circuit topologies are most adequate to use at ultra-low supply voltages. The choice of topology is based on various metrics, such as variation-resilience, delay, leakage power, total energy consumption, etc. An elaborate comparison between a number of topologies was therefore made, varying from very common logic families to much more exotic circuit topologies. As a result, preferred implementations for logic gates, inverters, latches and flip-flops have been proposed. Transmission Gate (TG) logic extended with nMOS stacking has been chosen as preferred topology for logic gates. Inverters are implemented as stacked nMOS inverters throughout this work. Different ratioless implementations of latches are investigated, which are used later on in the developed prototypes. Finally, the sizing of the gate-level building blocks which was used in these prototypes is summarized.

Chapter 4 analyzed various architecture-level dilemmas. It started with theoretical considerations on energy consumption, which were validated by experimental data and which provided more insight in the different mechanisms that influence the total energy consumption. The chapter continued with exploring the architectural consequences of using TG logic, especially when cascading multiple logic gates. The advantages and disadvantages of this cascading have been discussed, and an analysis which can be used to optimize the resulting trade-off has been presented. By cascading multiple logic gates, averaging of timing variations is obtained, which is very beneficial because of the high timing variations when operating at ultra-low supply voltages. Moreover, it was shown that the use of differential TG logic adds significantly to the variation-resilience of the system as well.

Subsequently, various pipelining schemes have been explored to assess their suitability for ultra-low-voltage designs. The conclusion was reached that latch-based pipelining is favorable over flip-flop-based pipelining due to the fact that it allows time borrowing. This time borrowing provided an extra measure to cope with the previously mentioned high timing variability. A simulation analysis was performed to study and quantify the effect of time borrowing. As a result, it has been shown that time borrowing is especially beneficial at ultra-low supply voltages and that the clock period can be drastically reduced by using a latch-based instead of a flip-flop-based pipeline. Furthermore, the chapter explained why deep pipelining is advantageous for ultra-low-voltage systems. To conclude, the design methodology which is employed for the different prototypes of this research was discussed profoundly.

Chapter 5 implements this design methodology in the first three ultra-low-voltage prototypes which have been implemented in this book. These three prototypes all consist of datapath blocks. Each design had different research purposes. The target of the first prototype, which was a 32-bit logarithmic adder fabricated in a 90 nm CMOS technology, was to confirm the robust operation of TG logic and latch-based deep pipelining in the ultra-low-voltage region. Measurement results have validated the proposed gate-level building blocks and architectural decisions.

From the experience of this first design, a second, more complex, prototype has been designed in the same technology: a 16-bit multiply-accumulate unit. The main changes which were executed are the use of differential TG logic, cascading multiple logic gates and using fully differential latches. Moreover, the MAC is a considerably larger datapath block which requires feedback. Measurements proved that these gate-level and architectural alterations with respect to the adder have improved the operating frequency, the energy consumption and the overall variation-resilience.

The aim of the third prototype was to study the influence of CMOS technology scaling. In order to perform this study, the 16-bit Multiply-Accumulate Unit (MAC) was redesigned in a 40 nm CMOS technology. The measurement results of the two MACs were extensively compared to reach conclusions on how technology scaling affects ultra-low-voltage systems. The scaling analysis of Chap. 2 has thus been extended with the analysis of this chapter which is based on the actual design and measurements of a large ultra-low-voltage circuit.

Chapter 6 then completed this research by presenting the fourth and final ultra-low-voltage prototype, which is a full JPEG encoder in the same 40 nm CMOS technology. The JPEG encoder has been chosen as a representative DSP block with which it is possible to demonstrate that the proposed design methodology is generally applicable in any large and complex ultra-low-voltage Digital Signal Processor (DSP) design. It has been designed by combining the theoretical studies, the gate-level explorations and the architecture-level investigations with the obtained insights from the designs of the previous three prototypes. The JPEG algorithm is explained to fully understand how the implementation of the different subblocks is necessary for correct JPEG encoding functionality. The design details of these subblocks are extensively covered, and the measures taken to increase energy-efficiency are presented. The book concludes with the successful measurements of this JPEG encoder which validate the proposed design methodology for any DSP design. Furthermore, improvements which could be implemented to reduce the leakage in the lookup tables and to increase their energy-efficiency are explored.

7.2 State-of-the-Art Comparison

This section will situate the obtained results of the different prototypes presented in this book in the current state-of-the-art literature, which was discussed in the beginning of this work in Sect. 1.5. Recall that two of the prototypes (the adder and the first version of the MAC) have been processed in a 90 nm CMOS technology, while the other two prototypes (the second version of the MAC and the JPEG encoder) have been fabricated in a 40 nm CMOS technology. In the Chaps. 5 and 6 which discussed these specific designs, the measurement results have been extensively compared to the most similar state-of-the-art designs. This section, on the other hand, will give a much more general overview by comparing the obtained results with all published measurement results of substantial digital circuits operating at supply voltages below 500 mV in CMOS technologies. An extensive table containing the details and the measured operating points of these papers can be found in Appendix A. The measurement data of the prototypes is added in diamond-shaped markers to the figures which were presented in Sect. 1.5.

Figure 7.1a shows the minimal functional supply voltage $V_{dd,min}$ as function of CMOS technology. As discussed before, $V_{dd,min}$ gives an idea of how variation-resilient a design is, since variability plays a larger role at lower supply voltages. It is therefore an interesting measure to compare various designs. However, since variability is very technology-dependent, $V_{dd,min}$ values should be compared within a specific technology node only. As can be seen in Fig. 7.1a, the four prototypes achieve the lowest $V_{dd,min}$ values of their technology nodes: the MAC is able to operate until the lowest $V_{dd,min}$ of 150 mV in the 90 nm technology node and the lowest $V_{dd,min}$ of 180 mV in the 40 nm node.

Figure 7.1b provides an overview of the supply voltages at which the MEP of the designs occurred. As already explained in Sect. 1.5, no pattern can be distinguished from this graph. It is apparent that the MEP occurs at very low supply voltages, however at which exact $V_{dd,MEP}$ it occurs is not so important.

A frequency comparison of all 40 nm and 90 nm published designs is shown in Fig. 7.2. The 90 nm prototypes of this work outperform the other publications in 90 nm CMOS technologies. For the 40 nm technologies, the MAC and the JPEG encoder perform similarly or better than the current state-of-the-art designs.

An overview of all frequency measurement points which were given in the publications of Appendix A is displayed in Fig. 7.3. The graph shows the results of all CMOS technology nodes in which designs have been fabricated. The

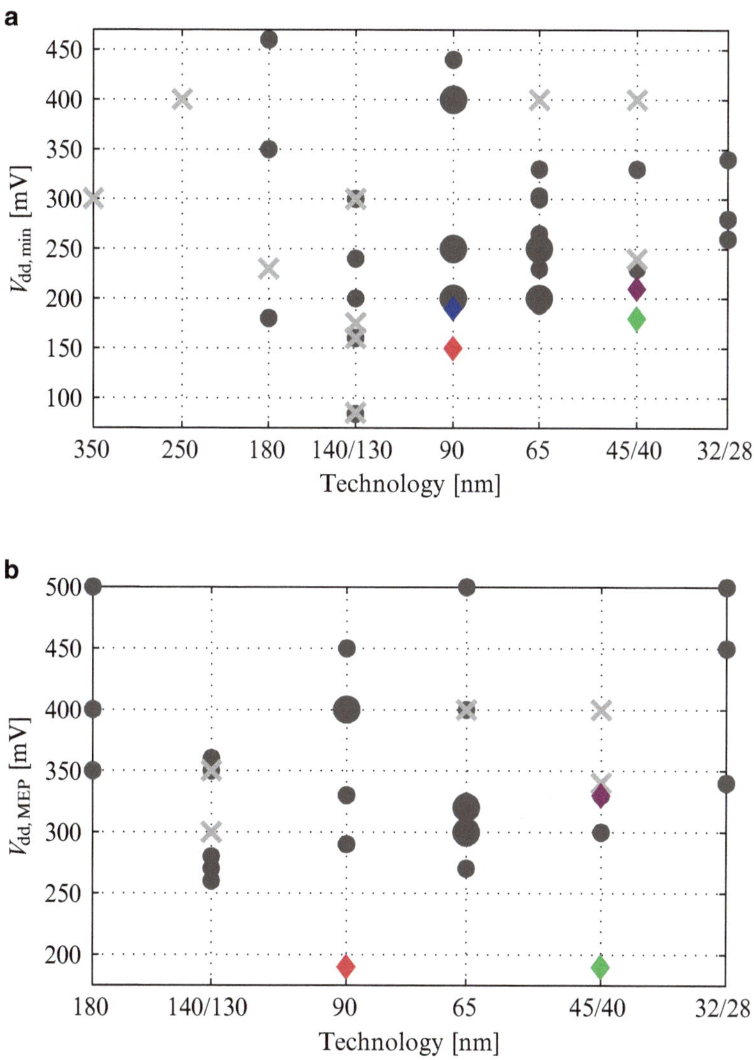

Fig. 7.1 Key voltages as function of CMOS technology node, with marker size proportional to population size: (**a**) minimal functional supply voltage $V_{dd,min}$ and (**b**) supply voltage at which the MEP occurs $V_{dd,MEP}$. A division is made between designs which use body biasing and those which do not. The *diamond-shaped markers* indicate the voltages of the four prototypes presented in this book

diamond-shaped markers indicate measurement results of the four prototypes. One of the aims of this book was to achieve operating frequencies well within the MHz-range to increase the industrial relevance of ultra-low-voltage digital circuits. Various techniques have been presented to not only increase the nominal speed

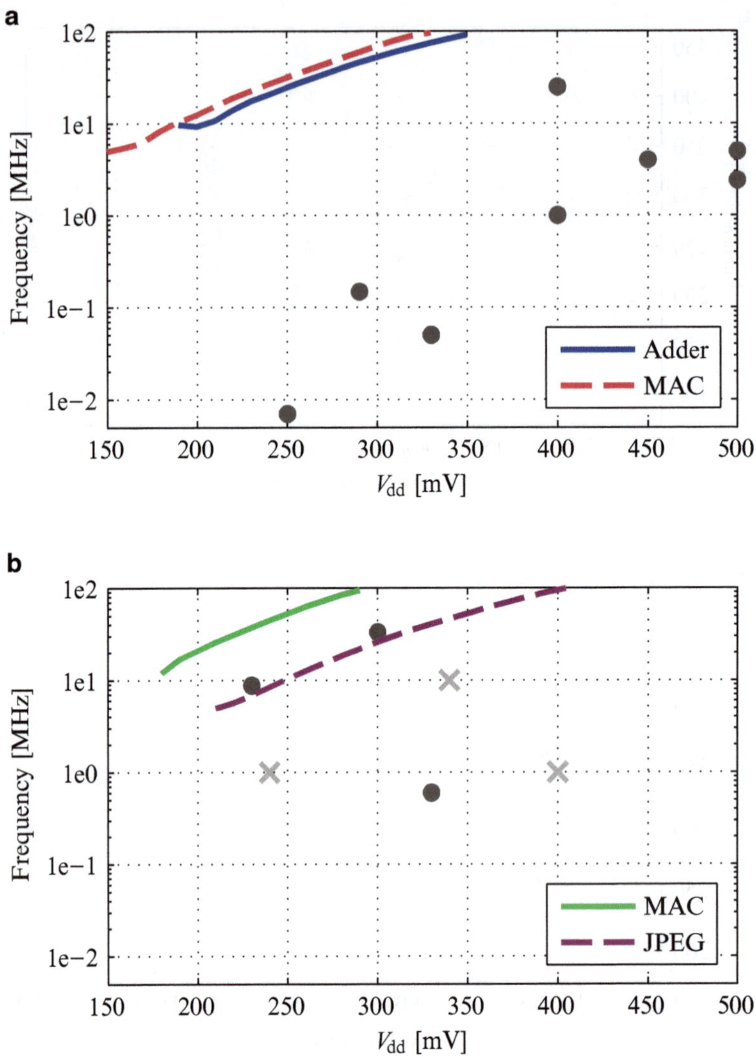

Fig. 7.2 Frequency as function of V_{dd} for all provided measurement points of the publications of Appendix A in specific technology nodes, including the mean measured operating frequencies of the four prototypes: (**a**) 90 nm CMOS and (**b**) 40 nm CMOS

of such circuits, but to cope with timing variations as well. As can be seen, the speed results of the presented prototypes are among the highest ever published. This has been established without compromising the energy-efficiency of the designs, as already extensively discussed and compared with similar designs in Chaps. 5 and 6.

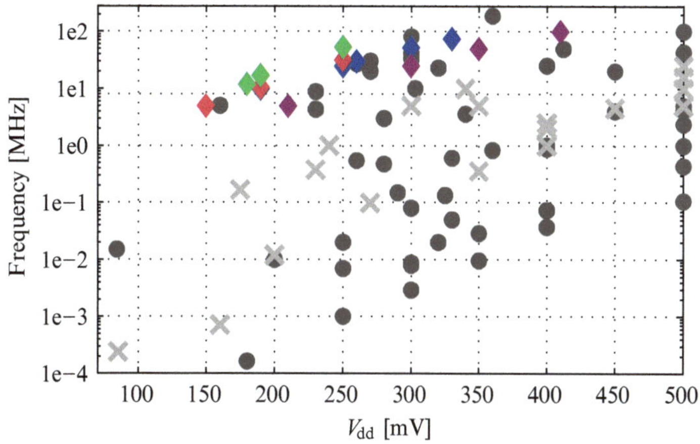

Fig. 7.3 Frequency as function of V_{dd} for all provided measurement points of the publications of Appendix A, for designs in all CMOS technology nodes. The *diamond-shaped markers* indicate values of the four prototypes presented in this book

7.3 Main Contributions

This work has realized the following contributions to the domain of energy-efficient, ultra-low-voltage digital circuit design:

- A comprehensive design methodology has been developed, covering all aspects of digital design: from gate-level design (e.g. the proposed use of differential TG logic with transistor stacking and differential ratioless latches with stacked nMOS inverters) up to architecture-level design (e.g. deep latch-based pipelining). This design strategy is generally applicable for all types of signal processing applications, as shown by the diverse prototypes implemented in the course of this research.
- Four innovative prototypes have been designed, fabricated and successfully measured in two CMOS technologies. These prototypes are functional at the lowest $V_{dd,min}$ values published of their technology nodes, when compared with state-of-the-art literature.
- The key target of ultra-low-voltage design, i.e. the energy-efficiency, of the prototypes has been validated by the measurement results and the state-of-the-art comparisons with similar designs.
- In order to increase the industrial relevance of ultra-low-voltage designs, relatively high speed performances well within the MHz range are required. All the presented prototypes achieve operating frequencies of $n \times 10$ MHz, which are among the highest published to date.
- A high yield is imperative to make the use of ultra-low-voltage circuits reliable, as these circuits are very sensitive to variations. This can be accomplished by guaranteeing a high variation-resilience of such circuits. However, this can only

be achieved by careful consideration of the impact of variations on all abstraction levels of digital design. This has been carried out in this research as follows: TG logic was preferred since it is inherently more robust than other circuit topologies, differential logic and latches have been shown to add to the total variation-resilience of a system, cascading of logic gates has been implemented so as to obtain averaging of the high timing variations, and latch-based pipelining has been employed because it allows time borrowing. The variation-resilience of the prototypes has been thoroughly demonstrated by the measurements.

- The proposed design methodology has not only been shown to deliver excellent results for the datapath and the control unit of a processor, but was also used to develop the implementation of the register tables of the JPEG encoder. Since the design target was to operate all subblocks of the JPEG encoder at the same supply voltage, the lookup tables were required to not only be functional at such low supplies, but to function at the same speed as the rest of the system as well. This has been accomplished by implementing them as register tables which were optimized for ultra-low-voltage operation with the same design methodology.
- The effects of technology scaling on ultra-low-voltage systems have been examined by using the MAC as a test vehicle. To the author's knowledge, this is the first time that measurement results of a full ultra-low-voltage digital system implemented in different CMOS technology nodes were presented to investigate the impact of scaling.
- Throughout this book, several theoretical and simulation analyses have been developed: a practical expression which estimates the minimum feasible supply voltage that can be expected for digital circuits in a certain CMOS technology node was proposed, a theoretical derivation of a variation factor was validated with measurement results, a simulation method to derive the maximally feasible logic depth has been developed and a simulation analysis has been established which quantifies the positive effect of time borrowing on the minimum clock period of a system.

7.4 Suggestions for Future Work

This research has concentrated on establishing and validating a complete ultra-low-voltage design methodology, all the way from transistor-level circuits up to architecture-level decisions. By using such a global design approach, it was possible to meet the goals of this work. In that sense, this book has paved the way for even more in-depth research, since many improvements can still be performed. Several suggestions for future work to accomplish these improvements are therefore listed below.

7.4.1 Energy-Efficient SRAM

The lookup tables in the JPEG encoder have been implemented as register tables which were able to operate at the same supply voltage and speed of the rest of the JPEG encoder. However, their contribution to leakage is quite high, as has been observed during measurements. Another possibility would be to use an energy-efficient SRAM, which, due to the fact that it is dominated by static energy, should be operated at a second, higher supply voltage. This could result in a significant improvement in leakage reduction. A case study has been explored regarding the inclusion of such an energy-efficient SRAM, which was extensively discussed in Sect. 6.7. Ideally, an on-chip implementation could be used to confirm the promising results of this simulated case study.

7.4.2 Other Technologies

The research presented in this work has been carried out for bulk CMOS technologies. Recently, several new process technologies have been developed, of which fully depleted SOI seems to be the most promising for ultra-low-voltage digital research. Its most important advantages include the near-ideal sub-threshold slope of its transistors, the reduced leakage and the decreased variability. These advantages are especially beneficial for advanced nanometer technologies with channel lengths below 40 nm. This is due to the increased leakage and variability which become specifically concerning for ultra-low-voltage designs below the 40 nm node. Promising results have already been established in e.g. Reyserhove et al [1], where body biasing has been utilized to adapt the Minimum-Energy Point (MEP) of the circuit to the desired workload. Although in this book, body biasing was shown to be an inefficient method for bulk CMOS technologies, it could play an interesting role in fully depleted SOI technologies, but this requires more research.

7.4.3 Standard Digital Design Flow

As explained in this book, the four prototypes have not been designed in the standard digital design flow. To make the proposed design methodology of this work attractive for industrial partners, it is imperative that it is incorporated in the standard design flow. Porting the layout generation to standard cell place and route tools should not pose a problem, as it was already generated by an automatic tool which used custom-made 'standard cells' for the gate-level building blocks. However, including the design strategy in digital synthesis might cause some larger issues. For example, timing verification tools do not yet standardly support latches, whereas this work has extensively demonstrated the advantages of latch-based pipelining for ultra-low-voltage systems. Therefore, extra research is necessary in this field to be able to demonstrate a functional system designed with the standard design flow.

7.4.4 *Inter-Die Variations*

Although the techniques proposed in this paper can cope with inter-die variations in the sense that they can guarantee robust functionality of the circuits in all process corners, global speed shifts due to process variations can currently not be avoided. From a commercial point of view, many systems have to be able to always function at the same speed, regardless of the circumstances. This could be realized by building a system that monitors the current speed of a system and adjusts the supply voltage accordingly to guarantee a certain, fixed performance.

7.4.5 *Temperature-Dependence*

The temperature-dependence of circuits operating at ultra-low supply voltages in environments around room temperature is rather limited, as discussed in Sect. 2.2.4. However, in broader temperature ranges, especially for temperatures below 0°C, the impact of temperature increases considerably and can have a detrimental effect on the functionality of ultra-low-voltage circuits. Therefore, it would be interesting to perform more research toward this temperature-dependence, for example by performing measurements on all dies at various fixed temperatures.

7.4.6 *Efficient DC-DC Converter*

Until now, this research has been conducted for stand-alone ultra-low-voltage circuits. If these blocks would be used in a much larger system, this system would likely make use of different voltage domains. Only the blocks which would really benefit from ultra-low-voltage operation would be functioning at such an ultra-low supply voltage. To avoid compromising the energy savings which are introduced by ultra-low-voltage operation, an efficient DC-DC converter that provides this low supply would need to be implemented in such a system.

Reference

1. Reyserhove H, Reynders N, Dehaene W (2014) Ultra-low voltage datapath blocks in 28 nm UTBB FD-SOI. In: Proceedings of the IEEE Asian solid-state circuits conference (A-SSCC), pp 49–52. DOI: 10.1109/ASSCC.2014.7008857

Appendix A
Current State-of-the-Art in Literature

This appendix provides an overview of the current state-of-the-art in literature on ultra-low-voltage digital circuit design in CMOS technologies. A subset of papers will be discussed.

The criteria for selection are the following:

- It must consist of a substantial digital circuit which has been fabricated and measured. Very simple digital circuits, such as a single logic gate or a ring oscillator, have been discarded. The implementations of the papers are included in the table, and range from adders and multipliers to full DSPs and microcontrollers.
- The designs must be fabricated in bulk CMOS technologies.
- The designs must be able to operate at ultra-low supply voltages. The requirement which has been used for this appendix is that they should be able to function at supply voltages below 500 mV.

Table A.1 contains all relevant details of the 41 selected papers and their measured operating points at supply voltages equal to or below 500 mV:

- The year in which the paper was published.
- Paper reference.
- A short description of the design.
- In which CMOS technology the design has been fabricated.
- Whether body biasing was employed or not.
- The following measured operating points:
  - At the minimal functional supply voltage $V_{\mathrm{dd,min}}$.
  - At the MEP (if provided).
  - Any other operating points (if provided).

All reported operating points were measured at room temperatures, except for the ones of papers 12, 21 and 22 (indicated by *), which were measured at 50 °C.

N. Reynders, W. Dehaene, *Ultra-Low-Voltage Design of Energy-Efficient Digital Circuits*, Analog Circuits and Signal Processing,
DOI 10.1007/978-3-319-16136-5

Table A.1 Overview of current state-of-the-art in literature on ultra-low-voltage digital circuit design

#	Year	Publication	What?	CMOS tech.	Body bias	Measured operating points $V_{dd,min}$	MEP	Other
1	2002	[27, 45]	MAC	140 nm	✓	175 mV, 166 kHz	–	–
2	2003/2001	[31, 46]	8 × 8 array multiplier	0.35 μm	✓	300 mV	–	–
3	2004/2005	[54, 55]	16-bit 1024-pt FFT	0.18 μm	✗	180 mV, 164 Hz	350 mV, 9.6 kHz	–
4	2005	[56]	32-bit RISC core	0.25 μm	✓	400 mV, 2 MHz	–	500 mV, 5 MHz
5	2005/2006	[6, 7]	32-bit Kogge-Stone adder	90 nm	✗	250 mV, 7 kHz	330 mV, 50 kHz	500 mV, 2.4 MHz
6	2006/2009	[59, 60]	Sensor processor	130 nm	✗	200 mV	360 mV, 833 kHz	–
7	2007	[57]	32-bit RISC core	0.18 μm	✓	230 mV, 375 kHz	–	350 mV, 5 MHz /500 mV,16 MHz
8	2007	[53]	Baseband processor	90 nm	✗	400 mV, 25 MHz	400 mV, 25 MHz	–
9	2007/2008	[16, 17]	Sensor processor	130 nm	✓	160 mV, 710 Hz	350 mV, 354 kHz	–
10	2007	[21]	8 × 8 FIR	130 nm	✓	85 mV, 240 Hz	–	200 mV, 12 kHz / 270 mV, 98 kHz
11	2008	[22]	DSP	90 nm	✗	440 mV	450 mV, 4 MHz	500 mV, 5 MHz
12*	2008/2009	[28, 29]	Motion estimation accelerator	65 nm	✗	230 mV, 4.3 MHz	320 mV, 23 MHz	–
13	2008/2009	[35, 36]	Microcontroller	65 nm	✗	300 mV, 8.7 kHz	500 mV, 434 kHz	–
14	2008/2009	[18, 51]	Sensor processor	0.18 μm	✗	460 mV	500 mV, 106 kHz	–
15	2009	[26]	Microcontroller	130 nm	✗	240 mV	280 mV, 475 kHz	–
16	2009/2010	[47, 48]	JPEG encoder	65 nm	✓	400 mV, 2.5 MHz	400 mV, 2.5 MHz	450 mV, 4.5 MHz /500 mV, 10 MHz
17	2009/2010	[42, 43]	14-tap 8-bit FIR	130 nm	✗	160 mV, 5 MHz	270 mV, 20 MHz	300 mV, 80 MHz / 360 mV, 187 MHz
18	2010	[12]	Frame decoder	40 nm	✗	330 mV, 600 kHz	330 mV, 600 kHz	–
19	2010	[9]	8-tap FIR	90 nm	✗	200 mV	290 mV, 148 kHz	–
20	2010	[10]	Sensor platform	0.18 μm	✗	350 mV	400 mV, 73 kHz	500 mV, 1 MHz

Table A.1 (continued)

#	Year	Publication	What?	CMOS tech.	Body bias	Measured operating points		
						$V_{dd,min}$	MEP	Other
21*	2010	[1]	CLB array	32 nm	✗	260 mV, 27 MHz	340 mV	–
22*	2010	[30]	SIMD accelerator engine	45 nm	✗	230 mV, 8.8 MHz	300 mV, 33 MHz	–
23	2010	[8, 50]	FPGA	90 nm	✗	200 mV	–	–
24	2010/2012	[2, 49]	Cardiac event detector	65 nm	✗	250 mV, 1 kHz	320 mV, 20 kHz	–
25	2011/2012	[38, 39]	8 × 8 multiplier	130 nm	✗	84 mV, 15.2 kHz	260 mV, 544.8 kHz	–
26	2011	[13]	Frame decoder	40 nm	✓	240 mV, 1 MHz	340 mV, 10 MHz	–
27	2011	[3, 20]	Biomedical signal processor	90 nm	✗	400 mV, 1 MHz	400 mV, 1 MHz	–
28	2011	[19]	AES coprocessor	65 nm	✗	193 mV	300 mV, 80 kHz	400 mV, 890 kHz
29	2011/2012	[15, 23]	DSP SoC	28 nm	✗	340 mV, 3.6 MHz	500 mV, 43.4 MHz	–
30	2011/2012	[25, 52]	16-bit 1024-pt FFT	65nm	✗	260 mV	270 mV, 30 MHz	300 mV, 41 MHz
31	2012	[33]	30-tap FIR	130 nm	✗	300 mV, 8 kHz	350 mV, 29 kHz	–
32	2012	[58]	FFT	65nm	✗	200 mV	450 mV, 20 MHz	–
33	2012	[37]	Neural signal processor	65 nm	✗	250 mV, 20 kHz	–	–
34	2012	[44]	8-bit microprocessor	65 nm	✗	300 mV, 2.95 kHz	300 mV, 2.95 kHz	400 mV, 37.2 kHz
35	2012	[24]	IA-32 processor	32 nm	✗	280 mV, 3 MHz	450 mV	500 mV, 100 MHz
36	2012/2013	[4, 5]	Microcontroller	65 nm	✗	265 mV	400 mV, 25 MHz	–
37	2012/2013	[40, 41]	32-bit RISC processor	65 nm	✗	200 mV, 10 kHz	325 mV, 133 kHz	500 mV, 4.2 MHz
38	2013	[11]	32-bit RISC core	130 nm	✓	300 mV, 5 MHz	300 mV, 5 MHz	500 mV, 25 MHz
39	2013	[34]	SoC for ExG	40 nm	✓	400 mV, 1 MHz	400 mV, 1 MHz	–
40	2014	[14]	32-bit processor	90 nm	✗	250 mV	–	–
41	2014	[32]	16-bit microprocessor	65 nm	✗	303 mV, 10 MHz	–	412 mV, 49 MHz

References

1. Agarwal A, Mathew S, Hsu S, Anders M, Kaul H, Sheikh F, Ramanarayanan R, Srinivasan S, Krishnamurthy R, Borkar S (2010) A 320mV-to-1.2V on-die fine-grained reconfigurable fabric for DSP/media accelerators in 32nm CMOS. In: Proceedings of the IEEE international solid-state circuits conference (ISSCC), pp 328–329. DOI:10.1109/ISSCC.2010.5433903
2. Akgun O, Rodrigues J, Leblebici Y, Owall V (2012) High-level energy estimation in the sub-Vt domain: Simulation and measurement of a cardiac event detector. IEEE Trans Biomed Circuits Syst 6(1):15–27. DOI:10.1109/TBCAS.2011.2157505
3. Ashouei M, Hulzink J, Konijnenburg M, Zhou J, Duarte F, Breeschoten A, Huisken J, Stuyt J, de Groot H, Barat F, David J, Van Ginderdeuren J (2011) A voltage-scalable biomedical signal processor running ECG using 13pJ/cycle at 1MHz and 0.4V. In: Proceedings of the IEEE international solid-state circuits conference (ISSCC), pp 332–334. DOI:10.1109/ISSCC.2011.5746341
4. Bol D, De Vos J, Hocquet C, Botman F, Durvaux F, Boyd S, Flandre D, Legat JD (2012) A 25MHz 7—W/MHz ultra-low-voltage microcontroller SoC in 65nm LP/GP CMOS for low-carbon wireless sensor nodes. In: Proceedings of the IEEE international solid-state circuits conference (ISSCC), pp 490–491. DOI:10.1109/ISSCC.2012.6177104
5. Bol D, De Vos J, Hocquet C, Botman F, Durvaux F, Boyd S, Flandre D, Legat JD (2013) Sleepwalker: A 25-MHz 0.4-V sub-mm^2 7—W/MHz microcontroller in 65-nm LP/GP CMOS for low-carbon wireless sensor nodes. IEEE J Solid State Circuits 48(1):20–32. DOI:10.1109/JSSC.2012.2218067
6. Calhoun B, Chandrakasan A (2005) Ultra-dynamic voltage scaling using sub-threshold operation and local voltage dithering in 90nm CMOS. In: Proceedings of the IEEE international solid-state circuits conference (ISSCC), pp 300–301. DOI:10.1109/ISSCC.2005.1493988
7. Calhoun B, Chandrakasan A (2006) Ultra-dynamic voltage scaling (UDVS) using sub-threshold operation and local voltage dithering. IEEE J Solid State Circuits 41(1):238–245. DOI:10.1109/JSSC.2005.859886
8. Calhoun B, Ryan J, Khanna S, Putic M, Lach J (2010) Flexible circuits and architectures for ultralow power. Proc IEEE 98(2):267–282. DOI:10.1109/JPROC.2009.2037211
9. Chang IJ, Park SP, Roy K (2010) Exploring asynchronous design techniques for process-tolerant and energy-efficient subthreshold operation. IEEE J Solid State Circuits 45(2):401–410. DOI:10.1109/JSSC.2009.2036764
10. Chen G, Fojtik M, Kim D, Fick D, Park J, Seok M, Chen MT, Foo Z, Sylvester D, Blaauw D (2010) Millimeter-scale nearly perpetual sensor system with stacked battery and solar cells. In: Proceedings of the IEEE international solid-state circuits conference (ISSCC), pp 288–289. DOI:10.1109/ISSCC.2010.5433921
11. Chen JS, Yeh C, Wang JS (2013) Self-super-cutoff power gating with state retention on a 0.3V 0.29fJ/cycle/gate 32b RISC core in 0.13—m CMOS. In: Proceedings of the IEEE international solid-state circuits conference (ISSCC), pp 426–427. DOI:10.1109/ISSCC.2013.6487799
12. Clerc S, Abouzeid F, Heinrich V, Jain A, Veggetti A, Crippa D, Roche P, Sicard G (2010) A 40nm CMOS, 1.27nJ, 330mV, 600kHz, bose chaudhuri hocquenghem 252 bits frame decoder. In: Proceedings of the IEEE international conference on IC design and technology (ICICDT), pp 78–81. DOI:10.1109/ICICDT.2010.5510284
13. Clerc S, Abouzeid F, Argoud F, Kumar A, Kumar R, Roche P (2011) A 240mV 1MHz, 340mV 10MHz, 40nm CMOS, 252 bits frame decoder using ultra-low voltage circuit design platform. In: Proceedings of the IEEE international conference on electronics, circuits and systems (ICECS), pp 117–120. DOI:10.1109/ICECS.2011.6122228
14. Craig K, Shakhsheer Y, Arrabi S, Khanna S, Lach J, Calhoun B (2014) A 32 b 90nm processor implementing panoptic DVS achieving energy efficient operation from sub-threshold to high performance. IEEE J Solid State Circuits 49(2):545–552. DOI:10.1109/JSSC.2013.2285384

15. Gammie G, Ickes N, Sinangil M, Rithe R, Gu J, Wang A, Mair H, Datla S, Rong B, Honnavara-Prasad S, Ho L, Baldwin G, Buss D, Chandrakasan A, Ko U (2011) A 28nm 0.6V low-power DSP for mobile applications. In: Proceedings of the IEEE international solid-state circuits conference (ISSCC), pp 132–133. DOI:10.1109/ISSCC.2011.5746251
16. Hanson S, Zhai B, Seok M, Cline B, Zhou K, Singhal M, Minuth M, Olson J, Nazhandali L, Austin T, Sylvester D, Blaauw D (2007) Performance and variability optimization strategies in a sub-200mV, 3.5pJ/inst, 11nW subthreshold processor. In: Proceedings of the IEEE symposium on VLSI circuits (VLSIC), pp 152–153. DOI:10.1109/VLSIC.2007.4342694
17. Hanson S, Zhai B, Seok M, Cline B, Zhou K, Singhal M, Minuth M, Olson J, Nazhandali L, Austin T, Sylvester D, Blaauw D (2008) Exploring variability and performance in a sub-200-mV processor. IEEE J Solid State Circuits 43(4):881–891. DOI:10.1109/JSSC.2008.917505
18. Hanson S, Seok M, Lin YS, Foo Z, Kim D, Lee Y, Liu N, Sylvester D, Blaauw D (2009) A low-voltage processor for sensing applications with picowatt standby mode. IEEE J Solid State Circuits 44(4):1145–1155. DOI:10.1109/JSSC.2009.2014205
19. Hocquet C, Kamel D, Regazzoni F, Legat JD, Flandre D, Bol D, Standaert FX (2011) Harvesting the potential of nano-CMOS for lightweight cryptography: an ultra-low-voltage 65 nm AES coprocessor for passive RFID tags. Springer J Cryptogr Eng 1(1):79–86
20. Hulzink J, Konijnenburg M, Ashouei M, Breeschoten A, Berset T, Huisken J, Stuyt J, de Groot H, Barat F, David J, Van Ginderdeuren J (2011) An ultra low energy biomedical signal processing system operating at near-threshold. IEEE Trans Biomed Circuits Syst 5(6):546–554. DOI:10.1109/TBCAS.2011.2176726
21. Hwang ME, Raychowdhury A, Kim K, Roy K (2007) A 85mV 40nW process-tolerant subthreshold 8x8 FIR filter in 130nm technology. In: Proceedings of the IEEE symposium on VLSI circuits (VLSIC), pp 154–155. DOI:10.1109/VLSIC.2007.4342695
22. Ickes N, Finchelstein D, Chandrakasan A (2008) A 10-pJ/instruction, 4-MIPS micropower DSP for sensor applications. In: Proceedings of the IEEE Asian solid-state circuits conference (A-SSCC), pp 289–292. DOI:10.1109/ASSCC.2008.4708784
23. Ickes N, Gammie G, Sinangil M, Rithe R, Gu J, Wang A, Mair H, Datla S, Rong B, Honnavara-Prasad S, Ho L, Baldwin G, Buss D, Chandrakasan A, Ko U (2012) A 28 nm 0.6 V low power DSP for mobile applications. IEEE J Solid State Circuits 47(1):35–46. DOI:10.1109/JSSC.2011.2169689
24. Jain S, Khare S, Yada S, Ambili V, Salihundam P, Ramani S, Muthukumar S, Srinivasan M, Kumar A, Gb S, Ramanarayanan R, Erraguntla V, Howard J, Vangal S, Dighe S, Ruhl G, Aseron P, Wilson H, Borkar N, De V, Borkar S (2012) A 280mV-to-1.2V wide-operating-range IA-32 processor in 32nm CMOS. In: Proceedings of the IEEE international solid-state circuits conference (ISSCC), pp 66–68. DOI:10.1109/ISSCC.2012.6176932
25. Jeon D, Seok M, Chakrabarti C, Blaauw D, Sylvester D (2012) A super-pipelined energy efficient subthreshold 240 MS/s FFT core in 65nm CMOS. IEEE J Solid State Circuits 47(1):23–34. DOI:10.1109/JSSC.2011.2169311
26. Jocke SC, Bolus J, Wooters S, Jurik A, Weaver A, Blalock T, Calhoun B (2009) A 2.6-—W sub-threshold mixed-signal ECG SoC. In: Proceedings of the IEEE symposium on VLSI circuits (VLSIC), pp 60–61
27. Kao J, Miyazaki M, Chandrakasan A (2002) A 175-mV multiply-accumulate unit using an adaptive supply voltage and body bias architecture. IEEE J Solid State Circuits 37(11):1545–1554. DOI:10.1109/JSSC.2002.803957
28. Kaul H, Anders M, Mathew S, Hsu S, Agarwal A, Krishnamurthy R, Borkar S (2008) A 320mV 56—W 411GOPS/Watt ultra-low voltage motion estimation accelerator in 65nm CMOS. In: Proceedings of the IEEE international solid-state circuits conference (ISSCC), pp 316–317. DOI:10.1109/ISSCC.2008.4523184
29. Kaul H, Anders M, Mathew S, Hsu S, Agarwal A, Krishnamurthy R, Borkar S (2009) A 320 mV 56 —W 411 GOPS/Watt ultra-low voltage motion estimation accelerator in 65 nm CMOS. IEEE J Solid State Circuits 44(1):107–114. DOI:10.1109/JSSC.2008.2007164

30. Kaul H, Anders M, Mathew S, Hsu S, Agarwal A, Krishnamurthy R, Borkar S (2010) A 300 mV 494GOPS/W reconfigurable dual-supply 4-way SIMD vector processing accelerator in 45 nm CMOS. IEEE J Solid State Circuits 45(1):95–102. DOI:10.1109/JSSC.2009.2031813
31. Kim C, Soeleman H, Roy K (2003) Ultra-low-power DLMS adaptive filter for hearing aid applications. IEEE Trans Very Large Scale Integration (VLSI) Syst 11(6):1058–1067. DOI:10.1109/TVLSI.2003.819573
32. Kim S, Seok M (2014) R-processor: 0.4V resilient processor with a voltage-scalable and low-overhead in-situ error detection and correction technique in 65nm CMOS. In: Proceedings of the IEEE symposium on VLSI circuits (VLSIC), pp 1–2. DOI:10.1109/VLSIC.2014.6858421
33. Klinefelter A, Zhang Y, Otis B, Calhoun B (2012) A programmable 34 nW/channel sub-threshold signal band power extractor on a body sensor node SoC. IEEE Trans Circuits Syst Express Briefs 59(12):937–941. DOI:10.1109/TCSII.2012.2231041
34. Konijnenburg M, Cho Y, Ashouei M, Gemmeke T, Kim C, Hulzink J, Stuyt J, Jung M, Huisken J, Ryu S, Kim J, de Groot H (2013) Reliable and energy-efficient 1MHz 0.4V dynamically reconfigurable SoC for ExG applications in 40nm LP CMOS. In: Proceedings of the IEEE international solid-state circuits conference (ISSCC), pp 430–431. DOI:10.1109/ISSCC.2013.6487801
35. Kwong J, Ramadass Y, Verma N, Koesler M, Huber K, Moormann H, Chandrakasan A (2008) A 65nm sub-Vt microcontroller with integrated SRAM and switched-capacitor DC-DC converter. In: Proceedings of the IEEE international solid-state circuits conference (ISSCC), pp 318–319. DOI:10.1109/ISSCC.2008.4523185
36. Kwong J, Ramadass Y, Verma N, Chandrakasan A (2009) A 65 nm sub-Vt microcontroller with integrated SRAM and switched capacitor DC-DC converter. IEEE J Solid State Circuits 44(1):115–126. DOI:10.1109/JSSC.2008.2007160
37. Liu TT, Rabaey J (2012) A 0.25V 460nW asynchronous neural signal processor with inherent leakage suppression. In: Proceedings of the IEEE symposium on VLSI circuits (VLSIC), pp 158–159. DOI:10.1109/VLSIC.2012.6243838
38. Lotze N, Manoli Y (2011) A 62mV 0.13—m CMOS standard-cell-based design technique using schmitt-trigger logic. In: Proceedings of the IEEE international solid-state circuits conference (ISSCC), pp 340–341. DOI:10.1109/ISSCC.2011.5746345
39. Lotze N, Manoli Y (2012) A 62 mV 0.13—m CMOS standard-cell-based design technique using schmitt-trigger logic. IEEE J Solid State Circuits 47(1):47–60. DOI:10.1109/JSSC.2011.2167777
40. Luetkemeier S, Jungeblut T, Porrmann M, Rueckert U (2012) A 200mV 32b subthreshold processor with adaptive supply voltage control. In: Proceedings of the IEEE international solid-state circuits conference (ISSCC), pp 484–485. DOI:10.1109/ISSCC.2012.6177101
41. Lutkemeier S, Jungeblut T, Berge H, Aunet S, Porrmann M, Ruckert U (2013) A 65nm 32b subthreshold processor with 9T multi-Vt SRAM and adaptive supply voltage control. IEEE J Solid State Circuits 48(1):8–19. DOI:10.1109/JSSC.2012.2220671
42. Ma WH, Kao J, Sathe V, Papaefthymiou M (2009) A 187MHz subthreshold-supply robust FIR filter with charge-recovery logic. In: Proceedings of the IEEE symposium on VLSI circuits (VLSIC), pp 202–203
43. Ma WH, Kao J, Sathe V, Papaefthymiou M (2010) 187 MHz subthreshold-supply charge-recovery FIR. IEEE J Solid State Circuits 45(4):793–803. DOI:10.1109/JSSC.2010.2042247
44. Makipää J, Turnquist MJ, Laulainen E, Koskinen L (2012) Timing-error detection design considerations in subthreshold: An 8-bit microprocessor in 65nm CMOS. J Low Power Electron Appl 2(2):180–196. DOI:10.3390/jlpea2020180
45. Miyazaki M, Kao J, Chandrakasan A (2002) A 175 mV multiply-accumulate unit using an adaptive supply voltage and body bias (ASB) architecture. In: Proceedings of the IEEE international solid-state circuits conference (ISSCC), pp 58–59. DOI:10.1109/ISSCC.2002.992937
46. Paul B, Soeleman H, Roy K (2001) An 8x8 sub-threshold digital CMOS carry save array multiplier. In: Proceedings of the IEEE European solid-state circuits conference (ESSCIRC), pp 377–380

47. Pu Y, Pineda de Gyvez J, Corporaal H, Ha Y (2009) An ultra-low-energy/frame multi-standard JPEG co-processor in 65nm CMOS with sub/near-threshold power supply. In: Proceedings of the IEEE international solid-state circuits conference (ISSCC), pp 146–147. DOI:10.1109/ISSCC.2009.4977350
48. Pu Y, Pineda de Gyvez J, Corporaal H, Ha Y (2010) An ultra-low-energy multi-standard JPEG co-processor in 65 nm CMOS with sub/near threshold supply voltage. IEEE J Solid State Circuits 45(3):668–680. DOI:10.1109/JSSC.2009.2039684
49. Rodrigues J, Akgun O, Öwall V (2010) A sub-1 pJ sub-VT cardiac event detector in 65 nm LL-HVT CMOS. In: Proceedings of the IEEE/IFIP VLSI system on chip conference (VLSI-SoC), pp 253–258. DOI:10.1109/VLSISOC.2010.5642669
50. Ryan J, Calhoun B (2010) A sub-threshold FPGA with low-swing dual-VDD interconnect in 90nm CMOS. In: Proceedings of the IEEE custom integrated circuits conference (CICC), pp 1–4. DOI:10.1109/CICC.2010.5617466
51. Seok M, Hanson S, Lin YS, Foo Z, Kim D, Lee Y, Liu N, Sylvester D, Blaauw D (2008) The phoenix processor: A 30pW platform for sensor applications. In: Proceedings of the IEEE symposium on VLSI circuits (VLSIC), pp 188–189. DOI:10.1109/VLSIC.2008.4586001
52. Seok M, Jeon D, Chakrabarti C, Blaauw D, Sylvester D (2011) A 0.27V 30MHz 17.7nJ/transform 1024-pt complex FFT core with super-pipelining. In: Proceedings of the IEEE international solid-state circuits conference (ISSCC), pp 342–343. DOI:10.1109/ISSCC.2011.5746346
53. Sze V, Chandrakasan A (2007) A 0.4-V UWB baseband processor. In: Proceedings of the ACM/IEEE international symposium on low power electronics and design (ISLPED), pp 262–267. DOI:10.1145/1283780.1283837
54. Wang A, Chandrakasan A (2004) A 180mV FFT processor using subthreshold circuit techniques. In: Proceedings of the IEEE international solid-state circuits conference (ISSCC), pp 292–293. DOI:10.1109/ISSCC.2004.1332709
55. Wang A, Chandrakasan A (2005) A 180-mV subthreshold FFT processor using a minimum energy design methodology. IEEE J Solid-State Circuits 40(1):310–319. DOI:10.1109/JSSC.2004.837945
56. Wang JS, Li HY, Yeh C, Chen TF (2005) Design techniques for single-low-Vdd CMOS systems. IEEE J Solid State Circuits 40(5):1157–1165. DOI:10.1109/JSSC.2005.845979
57. Wang JS, Chen JS, Wang YM, Yeh C (2007) A 230mV-to-500mV 375kHz-to-16MHz 32b RISC core in 0.18—m CMOS. In: Proceedings of the IEEE international solid-state circuits conference (ISSCC), pp 294–295. DOI:10.1109/ISSCC.2007.373410
58. Yang CH, Yu TH, Markovic D (2012) Power and area minimization of reconfigurable FFT processors: a 3GPP-LTE example. IEEE J Solid State Circuits 47(3):757–768. DOI:10.1109/JSSC.2011.2176163
59. Zhai B, Nazhandali L, Olson J, Reeves A, Minuth M, Helfand R, Pant S, Blaauw D, Austin T (2006) A 2.60pj/inst subthreshold sensor processor for optimal energy efficiency. In: Proceedings of the IEEE symposium on VLSI circuits (VLSIC), pp 154–155. DOI:10.1109/VLSIC.2006.1705356
60. Zhai B, Pant S, Nazhandali L, Hanson S, Olson J, Reeves A, Minuth M, Helfand R, Austin T, Sylvester D, Blaauw D (2009) Energy-efficient subthreshold processor design. IEEE Trans Very Large Scale Integr VLSI Syst 17(8):1127–1137. DOI:10.1109/TVLSI.2008.2007564

Index

A
adder, 114–120, 147
 architecture, 114
 elements, 114
 generate-propagate logic, 114
 logarithmic adder, 114
 measurement results, 117–119
 state-of-the-art comparison, 119
 tree adder, 114
adiabatic logic, 72–73
area, 51, 77
 active area, 117, 130, 159
averaging timing variations, 93, 123

B
Baugh-Wooley algorithm, 121
body biasing, 11, 13, 25–26, 58–59, 179
body effect, 25–26, 55, 58–59
Brent-Kung adder, 114

C
cascading logic, 92–98, 124
 realization, 98
channel length modulation, 22
complementary pass transistor logic, 63
CPL, *see* complementary pass transistor logic
current
 diffusion current, 23, 24
 drift current, 22, 24

D
DCT, *see* discrete cosine transform
design methodology, 107–109
DIBL, *see* drain-induced barrier lowering
digital signal processor, 10, 15, 120, 142
discrete cosine transform, 142
drain-induced barrier lowering, 26–27, 55
DSP, *see* digital signal processor
dynamic logic, 74–75

E
EDP, *see* energy-delay product
energy, 5–7, 86–92, 117, 134, 160
 dynamic energy, 5, 86
 static energy, 5, 86
energy-delay product, 7–8, 119, 134, 161
energy-efficiency, 1, 5, 10, 15, 160, 165–168

F
fall time, 50
FBB, *see* forward body biasing
figure of merit, 8, 119, 134
finite state machine, 146
flip-flop, 77, 80, 100
FOM, *see* figure of merit
forward body biasing, 25, 58
FSM, *see* finite state machine
full adder, 124

N. Reynders, W. Dehaene, *Ultra-Low-Voltage Design of Energy-Efficient Digital Circuits*, Analog Circuits and Signal Processing,
DOI 10.1007/978-3-319-16136-5

G
gain, 50
Gaussian distribution, 32, 93, 96

H
half adder, 124
Han-Carlson adder, 114, 147
high-K dielectric, 39
Huffman encoding, 143

I
I/O circuits, 109–111
inter-die variations, 31–32, 50, 180
intra-die variations, 32–33, 50
inverse narrow width effect, 68
inverter, 49, 101
  chosen topology, 76–77
  tristate inverter, 78–79
INWE, *see* inverse narrow width effect

J
joint photographic experts group, 142
JPEG, *see* joint photographic experts group
JPEG encoder, 141–169
  2D-DCT, 142, 146–148
  algorithm, 142–143
  Huffman encoding, 143, 149–155
  lookup tables, 143, 156–159, 165–168
  measurement results, 159–161
  quantization, 142, 148–149
  state-of-the-art comparison, 161–164
  timing, 144–145
  zigzag reordering, 143, 149

K
Kogge-Stone adder, 114

L
latch, 77–80, 100, 116, 125, 144
  differential, 79
  feedback latch, 121, 125–128
  ratioed, 78
  ratioless, 78
latency, 99, 116, 122
leakage, 30, 75, 106, 116, 118, 134, 161, 165
logic depth, 92
  optimal logic depth, 96
logic gate, 49
  cascading, 92–98
  chosen topology, 75–76
  differential logic, 96–97, 123
  ratioed, 60
  ratioless, 60
  regenerative, 50
  skewed, 49
  unskewed, 49, 50
lookup tables, 143, 156–159, 165–168

M
MAC, *see* multiply-accumulate unit
MC, *see* monte carlo
measurement setup, 109–111
memory elements, 77–80
MEP, *see* minimum-energy point
minimum-energy point, 5–7, 13, 106, 160–161, 174
mismatch, 33, 76, 158
Modified Baugh-Wooley algorithm, 121, 147
monte carlo, 32
multiply-accumulate unit, 36, 113, 120–138, 147
  architecture, 121–122
  elements, 124
  measurement results, 130–136
  state-of-the-art comparison, 136–137
  timing, 128–130

N
near-threshold, 8, 28, 165
noise margin, 49–50
non-overlapping clocks, 100–101, 116, 128, 144

O
oxide thickness, 38

P
pass transistor logic, 62–64
PDN, *see* pull-down network
Pelgrom coefficient, 40
performance, 2, 7, 16, 29–30
pipelining, 98–106
  benefits, 99–100
  concept, 98–99
  deep pipelining, 106, 116, 124, 144
  design considerations, 105–106
  drawbacks, 99–100
  flip-flop-based pipelining, 100
  latch-based pipelining, 100, 105, 116, 144

latency, 99
pipeline depth, 99
pipeline stage, 99
pipeline stage length, 99
throughput, 99
power, 4–5
power gating, 116–117, 119, 147, 157
predictive technology model, 36
process corners, 31, 51, 129
propagation delay, 29, 50
prototype
16-bit multiply-accumulate unit, 120–138
32-bit adder, 114–120
JPEG encoder, 141–169
pseudo-nMOS logic, 60–62
PTM, *see* predictive technology model
pull-down network, 48, 60
pull-up network, 48, 60
PUN, *see* pull-up network

R
race condition, 100
ratioed logic, 60
ratioless logic, 60
RBB, *see* reverse body biasing
regeneration, 50, 76, 92
reverse body biasing, 25, 58
reverse short-channel effect, 28, 58
rise time, 50
RSCE, *see* reverse short-channel effect

S
SAPTL, *see* sense amplifier-based pass transistor logic
scaling, 1–3, 11, 33, 35–42, 120, 132–136
constant electric field scaling, 2
constant voltage scaling, 2
SCE, *see* short-channel effect
sense amplifier-based pass transistor logic, 63
short-channel effect, 27–28
silicon-on-insulator, 11, 38, 179
SOI, *see* silicon-on-insulator
SRAM, *see* static random access memory
stacking effect, 55–56, 66, 77
standard CMOS logic, 48–60, 101, 129
static logic, 62
static random access memory, 156, 165–168, 179
strong inversion, 22
STSCL, *see* sub-threshold source-coupled logic
sub-threshold, 8, 23–24, 28
sub-threshold slope, 38–39, 87–88
sub-threshold source-coupled logic, 71–72
super-threshold operation, 29

T
technology
bulk CMOS, 14, 15, 38, 179
high-K dielectric, 39
high-performance process, 3, 42
low-leakage process, 3, 42
scaling, 35–42
temperature-dependence, 26, 34–35, 180
thermal voltage, 23
threshold voltage, 3, 24–28, 42, 48
body effect, 25–26
definition, 24–25
drain-induced barrier lowering, 26–27
short-channel effect, 27–28
temperature-dependence, 26
throughput, 99, 106
time borrowing, 101–105
transistor
HVT, 42, 166
LVT, 42, 123, 165
models, 41–42, 57
SVT, 42
type, 42–43
transmission gate, 64
transmission gate logic, 64–70, 92, 116
chosen topology, 75–76
differential TG logic, 96–97, 124, 144
logic functionality, 75–76
tristate inverter
full-CMOS implementation, 78
TG-based implementation, 78

U
ultra-low-voltage, 9, 15, 28, 40, 75, 99, 106, 163

V
variability, 31–33, 89–92
inter-die variations, 31–32
intra-die variations, 32–33

variation-resilience, 9, 12, 15, 31, 50, 75, 77, 80, 97, 108, 117, 124, 132, 142, 160, 174
velocity saturation, 2, 22–23
voltage transfer characteristic, 8, 49, 61
VTC, *see* voltage transfer characteristic

W
weak inversion, 8, 23–24

Y
yield, 9, 15, 31, 33, 95

MIX
Papier aus verantwortungsvollen Quellen
Paper from responsible sources
FSC® C105338

If you have any concerns about our products,
you can contact us on
ProductSafety@springernature.com

In case Publisher is established outside the EU,
the EU authorized representative is:
Springer Nature Customer Service Center GmbH
Europaplatz 3, 69115 Heidelberg, Germany

Printed by Libri Plureos GmbH
in Hamburg, Germany